Quality of Life

Information Technology, Management and Operations Research Practices

Series Editors:
Vijender Kumar Solanki, Sandhya Makkar, and Shivani Agarwal

This new book series will encompass theoretical and applied books and will be aimed at researchers, doctoral students, and industry practitioners to help in solving real-world problems. The books will help in the various paradigms of management and operations. The books will discuss the concepts and emerging trends on society and businesses. The focus is to collate the recent advances in the field and take the readers on a journey that begins with understanding the buzz words like employee engagement, employer branding, mathematics, operations, technology, and how they can be applied in various aspects. It walks readers through engaging with policy formulation, business management, and sustainable development through technological advances. It will provide a comprehensive discussion on the challenges, limitations, and solutions of everyday problems like how to use operations, management, and technology to understand the value-based education system, health and global warming, and real-time business challenges. The book series will bring together some of the top experts in the field throughout the world who will contribute their knowledge regarding different formulations and models. The aim is to provide the concepts of related technologies and novel findings to an audience that incorporates specialists, researchers, graduate students, designers, experts, and engineers who are occupied with research in technology, operations, and management-related issues.

Performance Management
Happiness and Keeping Pace with Technology
Edited by Madhu Arora, Poonam Khurana, and Sonam Choiden

Soft Computing Applications and Techniques in Healthcare
Edited by Ashish Mishra, G. Suseendran, and Trung-Nghia Phung

Employer Branding for Competitive Advantage
Models and Implementation Strategies
Edited by Geeta Rana, Shivani Agarwal, and Ravindra Sharma

Analytics in Finance and Risk Management
Edited by Sweta Agarwal, Nidhi Malhotra, and T. P. Ghosh

Quality of Life
An Interdisciplinary Perspective
Edited by Shruti Tripathi, Rashmi Rai, and Ingrid Van Rompay-Bartels

For more information about this series, please visit: https://www.routledge.com/Information-Technology-Management-and-Operations-Research-Practices/book-series/CRCITMORP

Quality of Life
An Interdisciplinary Perspective

Edited by
Shruti Tripathi, Rashmi Rai, and Ingrid Van Rompay-Bartels

CRC Press
Taylor & Francis Group
Boca Raton London New York

CRC Press is an imprint of the
Taylor & Francis Group, an **informa** business

First edition published 2022
by CRC Press
6000 Broken Sound Parkway NW, Suite 300, Boca Raton, FL 33487-2742

and by CRC Press
2 Park Square, Milton Park, Abingdon, Oxon, OX14 4RN

© 2022 Taylor & Francis Group, LLC

CRC Press is an imprint of Taylor & Francis Group, LLC

Library of Congress Cataloging-in-Publication Data
Names: Tripathi, Shruti, editor. | Rai, Rashmi, editor. |
Van Rompay-Bartels, Ingrid, editor.
Title: Quality of life: an interdisciplinary perspective / edited by
Shruti Tripathi, Rashmi Rai and Ingrid Van Rompay-Bartels.
Description: First edition. | Boca Raton, FL: CRC Press, [2021] |
Series: Information technology, management and operations research practices |
Includes bibliographical references and index.
Identifiers: LCCN 2021006760 (print) | LCCN 2021006761 (ebook) |
ISBN 9780367433994 (hbk) | ISBN 9781032048307 (pbk) |
ISBN 9781003009139 (ebk)
Subjects: LCSH: Quality of life.
Classification: LCC HN25 .Q333 2021 (print) |
LCC HN25 (ebook) | DDC 306—dc23
LC record available at https://lccn.loc.gov/2021006760
LC ebook record available at https://lccn.loc.gov/2021006761

ISBN: 978-0-367-43399-4 (hbk)
ISBN: 978-1-032-04830-7 (pbk)
ISBN: 978-1-003-00913-9 (ebk)

Typeset in Times
by codeMantra

Contents

Foreword

It is owing to the wonders of modern technology that Dr. Shruti Tripathi and I came into contact through our connection over LinkedIn.com. As we believe in our culture, perhaps it could be due to our karmic connections that Professor Tripathi and I eventually ended up discussing the topic of "Quality of Life: An Interdisciplinary Perspective", a concept that is very close to my heart. Initially, I had been asked to contribute an article on Quality of Life by intertwining it with Bhutan's development philosophy of Gross National Happiness (GNH). However, I had to regretfully decline the offer as it came at a time when I was fully engaged with my committee assignments and winter parliamentary session, and hence, I decided to make up for my failure by agreeing to write the Foreword for this book.

Although I do not have any personal acquaintance with authors of the chapters, going by their curriculum vitae I am thoroughly convinced that they are highly intellectual and vastly experienced individuals with different backgrounds. It is a privilege to learn that the chapters in this book are contributed by authors who have a vivid understanding of what Quality of Life entails and how to work toward improving it. The advantage of having multiple authors from various parts of the world is that each one of them will bring diverse socioeconomic, political and cultural, perhaps even spiritual perspectives on their topics of choice. Given their academic qualifications and nature of their professions and expertise, the topics have been very aptly chosen.

I am aware that I would not do justice in making comments that will be commensurate with the contents of the chapters. Nonetheless, as I live in a country that places tremendous importance on promoting happiness of the people by emphasizing on not only individual but collective happiness, I do stand a ground that gives me some confidence to say what I feel appropriate about this book, of course with much hesitation.

The ultimate goal of our life, no matter where we live, is happiness; to live a quality life, be it in terms of our health, work, social relationships, emotional well-being, income, etc. Keeping this as the background, in Bhutan, our His Majesty, the fourth King, propounded the philosophy of GNH in 1972. His Majesty profoundly felt that although GDP is important, it will not necessarily bring happiness to the citizens. Therefore, the GNH philosophy stresses on ensuring a balance between the material needs of an individual and his/her mental, spiritual and emotional state within a stable environment.

Every individual aspires to live a quality life, life filled with joy and happiness. However, the methods that each of us adopts to achieve happiness are different. Some even think that they would attain happiness by indulging in violence. It is perhaps with this understanding that the project on "Quality of Life: Interdisciplinary Perspective" was conceptualized.

From discussing the appropriateness of assessing Quality of Life using human development index to improving Quality of Life in nursing homes in Europe, this book deals with subject matters that discuss improving Quality of Life through global citizenship education and technological innovation for environmental sustainability and Quality of Life. It contains evidence-based case studies on workers' well-being and workplace environment in some European countries, international students' well-being achievements through the lens of the capability approach and spatial assessment of Quality of Life using composite index of an Indian state Kerala.

As the title suggests, this book discusses a range of topics that are very relevant in provoking a rethinking in our ways of living. It discusses key factors that determine work–life balance and its implication on Quality of Life, besides discussing integrating human well-being and Quality of Life derived from nature-based livelihoods in a modern economy and nurturing resilience and Quality of Life. Considering the modern way of living, this book also has a chapter on the role of smart cities in enhancing Quality of Life.

In this day and age, when blind consumerism and modern decadence influence us to believe in a material pursuit to achieve happiness, not all hope is lost when individuals such as the authors in this book came together to show us alternative paths to promote quality life. Often, we develop a propensity to negate our mental and physical well-being in our quest for greedy accumulation of wealth.

The book on "Quality of Life" covering various disciplines therefore comes as a gentle reminder to rethink and restrategize our approach to life and our ways of living harmoniously with humans, other sentient beings and the planet we live on. The discussions in this book are ever more relevant today than before, and I can vouch that this book can serve as a guide for the readers to practice a meaningful living of our precious lives.

You will not regret having an individual copy – it is certainly an interesting read!

Nima
Member of Parliament
National Council of Bhutan

Preface

The year 2020 is a landmark in human history. It was a year of human valour and conviction overcoming any adversity of life. It is a coincidence that in the same year, we started working on such an important topic, "Quality of Life", exploring it in various ways through academic contributions across the globe. Despite the upheavals, anxiety and the grave situation pertaining to COVID-19 affecting all nations alike, our academic contributors and we as editors remained resilient.

Since 2018, our editorial team has collaborated on different international research projects focusing on sustainable development and global citizenship, in which the concept of "Quality of Life" is one of the fundamental criteria. Our thoughts were always on similar lines to explore more in the field of Quality of Life which formally took shape in the form of the present book. Quality of Life being a diverse and vibrant field helped us to get a highly encouraging response from academicians and practitioners working in and across various fields, and from varied cultural and national backgrounds, to come together and embrace the topic "Quality of Life: An Interdisciplinary Perspective" in a book. This book reflects our ambition to explore interdisciplinary studies that bring together the way individuals, communities, and societies are thinking, reacting, and transforming our organizations, business models, and landscapes locally and globally in search of Quality of Life.

For ages, the human race has always looked for opportunities to better themselves and the situations they live in. The basic idea behind this is to boost their well-being and ultimately enrich their happiness which they measure in terms of health, wealth, comfort, happiness, security, safety, etc. of individuals and societies all together.

Quality of Life measures the general well-being of individuals and societies in a holistic way to measure the objective as well as subjective interests and happiness of all. It is also the perception an individual holds related to the position of its life under the umbrella of social, cultural, and different value systems through which they pursue their passions and ambitions. Despite being inherently ambiguous, it provides a thrust toward a better life that supplements the people's understanding of themselves and the world.

For people living in urban areas, smart city development can improve the quality of their lives from various aspects. Now the question arises how the Quality of Life and smart city development are related? With advancement in technology improving various issues such as traffic congestion, waste management, disaster prediction through early warning signals, etc., the lives of residents will be much easier and more comfortable. Smart cities will increase the purchasing power of residents by providing technological solutions to the majority of their issues, which will ultimately result in improved Quality of Life. "Quality of Life" also seeks to measure how secure residents feel while moving around the city, how much time is wasted in traffic; pollution and crime rate; identifies the shortest possible commute length, thereby making transportation faster and cheaper; and measures the cost of living. Such inclusive interventions will not only better the Quality of Life but also improve the economic condition of the city.

The concept of the smart city encompasses various factors such as smooth flow and delivery of sustainable sources of energy, digitized infrastructure, planned roadways, smart housing, and above all, better opportunities for education.

Education does not simply mean learning alphabets and science; rather, it is a resourceful tool that teaches us to be humans first and create opportunities that enable us to thrive outside the restricted world of our own inhibitions and impediments. Intrinsic education transcends cultural and geographical boundaries, making people more receptive to the environment. It is the multidimensional stratagem that affects physical health, emotional well-being, personal and future safety, and ultimately culminates to the development of a diverse aware and inclusive community.

COVID-19 taught us the importance of health. The field of Quality of Life refers to it as Health-Related Quality of Life (HRQOL). It encompasses mental, physical, emotional health along with the social functioning capability of an individual. Earlier, the term "Health" was viewed as the impact of incurable diseases, research related to treatments, and classification of short-term and long-term disabilities. But over the years (since the 1980s), health as a measure of Quality of Life has evolved with respect to technological advancement in medical science. In 2017, the term "QUIMP" was recognized in medicine, which indicates the impairment of Quality of Life. Few unique methods have been developed to measure health in terms of Quality of Life.

Successful ageing is yet another aspect of health-related Quality of Life as it aims to improve the life of elderly people. A society's responsibility is not restricted to formulating support policies for elderly people but also to make the natural process of ageing much smoother. The field of Biomedical Sciences has defined the term "successful ageing" as improving life expectancy and at the same time trying to minimize any mental or physical disability. Low Quality of Life in old age can be attributed to morbidity and functional constraints. Successful ageing can take place if people with diseases can also lead successful lives. Environment plays a very important role in achieving the target of a successful life.

Every individual aspires for well-being while interacting with the environment, be it the social or the natural environment. Economies operate and flourish by consuming natural resources such as water, timber, fisheries, and plants, and access to these environmental services and amenities, which is resulting in imbalanced growth. An unsullied environment provides satisfaction which is therapeutic and a remedy to everyday stress. Green spaces are a source of satisfaction that help us get going with our physical activities.

Clean air, land and water are fundamental to any community life. From villages to cities, the different population groups have formed a complex relationship with their daily living environment. It is not limited to residential conditions, the availability and reach of services, noise level and other tangible must-haves, but expands to the coexistence with nature. Therefore, multidimensional Quality of Life measurements are needed to study this human–environment relationship. Any depletion will have negative consequences in the short-term as well as long-term well-being. This suggests that Quality of Life is not solely hinged on physical and social quality of environment, but also on how individuals intertwine their ideologies and cultures within the framework of their inhabitations.

All this shows that bringing quality in life is a process. It is a seed that has to be sown in the correct soil, nurtured with right nutrients and most importantly taken care of so that it may eventually grow into an enormous tree which is deeply rooted and has a strong foundation which stands tall even in a storm. It is a marathon in which the main aim is to finish the race and to reach the goal. Quality takes time to build and especially when associated with life has to be accurately defined with all the parameters clearly stated, leaving no ambiguity in the minds of people. To walk towards the sphere of vitality requires consistency in actions such that each effort combines with that of the community and produces a synergy to enhance the overall Quality of Life.

This project has been a learning experience for all of us as researchers and editors. It made us discover that geographical boundaries are more about maps than about our minds. In an early response to this book, we received 106 chapter proposals but only 11 reached the final stage. The chapters are from leading universities and institutions which have contributed significantly to the field of Quality of Life. The different chapters not only contribute to the existing body of research through their new insights and analyses but will also propel readers and policy makers to explore this area more.

Best wishes,

—Dr. Shruti Tripathi
Associate Professor, HR & OB, Amity University, Noida, India,

—Dr. Rashmi Rai
Assistant Professor, School of Business and Management,
CHRIST (Deemed to University), Bangalore, India,

—Dr. Ingrid Van Rompay-Bartels
Senior Researcher & Lecturer at International School of
Business at HAN University of Applied Sciences, the Netherlands.

Acknowledgements

This book would not have been possible without the contribution and support of many wonderful people across the world. Completing this book was harder than we thought and more rewarding than we could have ever imagined, and in this journey, we realized that the world is a better place, where you have many people who want to develop and lead others. What makes it even better are the people who want to share the gift of their time to mentor others. Thanks to everyone who strive to grow and help others grow, and on this note, we would like to express our sincere gratitude to Professor Dr. Linley Lord, Sónia Rosana Alves de Brito, Dr. Eugénia de Matos Pedro, Professor Dr. Andrew Kakabadse, Dr. Aarti Shyamsunder, Dr. Carlos Vargas Tamez, Dr. Tetyana Vereschchahina, Prof (Dr.) Sarmistha Sarma, Dr. Varada Nikalje, Dr. Amruta Jajoo, Dr. Ashima Saxena, Kelly Hoareau (nee Bucas), Dominique Benzaken, Md. Uzir Hossain (Uzir), Hussam Al Halbusi, Chetan Vaidya, Prof. Joy Sen, K.C. Lalmalsawmzauva, Joost van Hoof, Dr. Oscar Zanutto, and at last our two dear students Divyanshi Singh and Mahima Esther Obed who helped us in every possible way and last but not the least our families who supported us in this beautiful journey.

Editors

Shruti Tripathi, PhD, is an Associate Professor in the area of HRM & Organizational Behaviour and Former HOD at Amity International Business School, Amity University, Noida, India. She earned her PhD in the area of Quality of Life from University of Allahabad, one of the oldest and top-ranked universities of India. She is NET-qualified and has a first-class academic career. Dr. Tripathi's research interests include Quality of Life, sustainable development, quality of work life, future of work, the emerging issues of management and technology especially in the post-pandemic world, changes in societal values on career decisions, etc. Her research contributions have been accepted in many international and national journals of repute.

Rashmi Rai, MBA, PhD, is an Assistant Professor in School of Business and Management, CHRIST (Deemed to be University). She is also the coordinator of MBA (International Business) Program at her university. Dr. Rai's main areas of interest are the emerging issues of Human Resource Management and Organizational Behaviour. She also has a keen interest in research that includes varied topics such as Quality of Work Life, successful ageing and work–life balance.

Ingrid Van Rompay-Bartels, MA, PhD, earned her degrees from Leiden University, the Netherlands. She is a Senior Lecturer in Intercultural Learning at International School of Business at HAN University of Applied Sciences. As a Senior Researcher, Dr. Van Rompay-Bartels also works for the Research Centre of International Business focusing on global citizenship, intercultural learning and competence, diversity and inclusion, and intercultural (virtual) collaboration competence.

Contributors

Luigi Aldieri, PhD, is an Associate Professor of Economics at the University of Salerno. He earned a PhD in Sciences Economiques et de gestion at the Solvay Business School of Economics and Management (SBS-EM) of Université Libre de Bruxelles (ULB). He earned his PhD in Economics at the University of Naples Federico II. Dr. Aldieri's research interests embrace applied econometrics, the measurement of knowledge, geographic spillovers and economic performance of large international firms. He has published several research articles and is a member of the editorial board of reputed journals.

Icy F. Anabo is a PhD candidate at the University of Deusto, Spain, and earned a Joint Master's Degree in Lifelong Learning: Policy and Management from the UCL Institute of Education and the University of Deusto. Her research interests include internationalization, the capability approach's applications in program evaluation, ecological perspectives in higher education, and research ethics.

Ole Petter Anfinsen, MBA, MSc, is the CEO of AEHP and a board member of EDBAC. He earned his degrees from Henley Business School – where he is also completing a Doctor of Business Administration degree focusing on Senior Executive Health and Performance.

Laura Annella, PhD, is a geriatric psychologist, with her degree in Cognitive Neuroscience and is specializing in body psychotherapy. She works in care facilities for dependent seniors managed by CADIAI Cooperativa Sociale developing projects and interventions to improve the Quality of Life of residents, support family caregivers, train and supervise care workers.

Nithin Babu, MSc, completed his undergraduate degree in Civil Engineering and Master's degree in Urban Planning. He works as a Research Associate at NIT Calicut. He is involved in research on the topics of low-cost housing for flood affected tribal communities and culturotope mapping as a planning tool in an urban cultural landscape.

James Baker currently works at Jaguar Land Rover and serves as Global Head of Talent Acquisition. His role includes TA strategy and operations for blue collar, white collar, temporary staffing, early careers and leadership recruitment up to executive board level. James is also a part-time research associate at Henley Business School.

Dr. Gouri Sankar Bhunia is currently involved as a GIS Expert in various smart city implementation programmes in India. His research interests include urban planning, infrastructure planning and mapping, environmental modelling, risk assessment, natural resources mapping and modelling, data mining and information retrieval using geospatial technology. Dr. Bhunia was awarded the Senior Research Fellowship Grant from Indian Council of Medical Research.

John Bourke, MBA, MSc, is the President of The Business Excellence Institute. He has worked in a range of industries and countries, founded three businesses, and consulted for governments and companies such as Google. He earned a BSc in Applied Physics, an MBA, and an MSc in Business and Management Research.

Belinda Bramley is an Associate at NLA International, a company creating concepts and projects to provide sustainable social, environmental, and economic benefits in the marine and maritime environments. Prior to that, she was Head of Impact at Nekton, an NGO undertaking deep ocean science in partnership with Indian Ocean countries to develop local capacity and accelerate sustainable ocean governance. With Angelique, she co-developed a joint SeyCCAT Nekton Deep Blue Grants scheme, which enabled six Seychellois citizens to undertake deep sea research as part of a six-week expedition in Seychelles waters and two of these to undertake visiting marine science fellowships at Oxford University. Belinda is a chartered accountant and passionate environmentalist, and served as CFO for Earthwatch Europe and the Global Canopy Programme.

Dr. Uday Chatterjee is an Assistant Professor in the Department of Geography, Bhatter College, Dantan, Paschim Medinipur, West Bengal, India. His areas of research interest cover urban planning, social and human geography, applied geomorphology, hazards & disasters, environmental issues, land use and rural development. Dr. Chatterjee has contributed various research papers published in several reputed national and international journals and edited book volumes. He has served as a reviewer for many international journals.

Iciar Elexpuru-Albizuri, PhD, is a Full Professor at the Department of Education at the Faculty of Psychology and Education, University of Deusto, Spain, where she teaches developmental psychology and ethics. She is a member of the eDucaR team, accredited as excellent. Dr. Elexpuru-Albizuri's research work focuses on values development and lifelong learning.

Dr. Mohammed Firoz C. earned a PhD degree from IIT Kharagpur and works as an Assistant Professor at NIT Calicut, India, where he has been involved in teaching, researching, and consulting since July 2004. Dr. Firoz's fields of interest includes rural urban interface studies, sustainable urbanism, regional development and Quality of Life studies.

Javier Ganzarain, MSc, is the Co-Founder of AFEdemy, and earned a Computer Science degree and a Master of Science (MSc). He has gained extensive international experience in the management of various projects, ICT and Healthy Ageing areas while working in Germany, Belgium, the Netherlands and Spain. His research areas of interest include human centred innovation, design and creativity.

Dr. Eglé Gerulaitiene is an Associate Professor at the Institute of Education, Vytautas Magnus University, and Head of the Project Management Unit in the Department of Research and Innovation. Her research interests include intercultural education, diversity and multiculturalism, implementation of innovations in the curriculum, application of innovative methods in the educational process, and development of intercultural competence.

Ângela Gonçalves earned an MSc in Biomedical Sciences, University of Beira Interior (UBI). Gonçalves is research fellow of the BIO-ALL project (Biohealth Gear Box Alliance) at UBImedical – UBI's incubator for the areas of health and well-being, and researcher at NECE, Research Center in Business Sciences, UBI.

Upeksha Hettiarachchi is a Master's student at Duke University with an academic focus in environmental economics, environmental governance, and ocean policy. She has experience in developing models for conservation and climate smart agriculture within local communities and carbon offsetting for the private sector. With a foot in both island nations of Seychelles and Sri Lanka, she has a vested interest in ensuring a sustainable relationship between people and the ocean.

Aditi Joshi, MBA, earned her Bachelor's in Commerce (DeenDayal Upadhyay Gorakhpur University) and then her MBA in Human Resources from Birla Institute of Management (BIT, Mesra). She has experience both in academia as faculty in Madan Mohan Malviya University of Technology and an other management college and has worked in industry as well. She is currently working with Gradestack Learning Pvt. Ltd as Associate Manager.

Parth Joshi is a development, research, and advocacy professional based in New Delhi with more than 12 years' experience in project management, strategic research, public policy, and communication. He has experience of working in a wide range of sectors with industry, academia, government, and local communities towards creating sustainable and inclusive growth models through grassroots projects, academic discourses, and industry forums.

João Leitão, PhD, earned his degree in Economics, University of Beira Interior (UBI); Habilitation in Technological Change and Entrepreneurship, IST, University of Lisbon (UL), Portugal. Dr. Leitão is an Associate Professor in habilitation, UBI; Director of the UBIExecutive, Business School; research fellow at the NECE, Research Center in Business Sciences, UBI; associate researcher of the CEG-IST – Centre for Management Studies of Instituto Superior Técnico & of the ICS – Instituto de Ciências Sociais, UL. His research interests include entrepreneurship, innovation and organizational economics.

Dinabandhu Mahata is a Researcher at the Department of Geography, Central University of Tamil Nadu. His work mainly focuses on Urban Geography, and links with the environment. He has written journal articles and book chapters. He received a Research Grant from the International Union for the Scientific Study of Population (IUSSP) for International Population Conference in Cape Town, South Africa.

Rasa Naujaniene is an Associate Professor at the Social Work Department of Vytautas Magnus University (Lithuania). In her academic career, she is an author of several publications in social gerontology and gerontological social work field as well. She is a participant in several research projects in this area. As a professional supervisor, she is well acquainted with the field of social care for the elderly.

Dina Pereira, PhD, earned her degree in Industrial Management and Engineering, University of Beira Interior (UBI); Manager of UBImedical – UBI's incubator for the areas of health and well-being; Researcher at NECE, Research Center in Business Sciences, UBI; Researcher at Centre for Management Studies of Instituto Superior Técnico (CEG-IST), University of Lisbon.

Angelique Pouponneau is CEO of the Seychelles' Conservation and Climate Adaptation Trust (SeyCCAT). Angelique is a Barrister and Attorney at law, a legal expert in climate change, ocean and global commons, an international speaker on youth, Blue Economy, islanders' rights in the face of rising sea levels, climate change and legal barriers to gender equality. She is a Queen's Young Leader, experienced in running not-for-profit organizations and working in multi-stakeholder contexts. Angelique has a track record of bringing about impact for local communities, especially youth and women.

Dr. Ashita Raveendran is presently working with the National Council of Educational Research & Training (NCERT), prior to which she was teaching economics at the graduate and postgraduate level. She completed her doctoral studies in Industrial Economics at Mahatma Gandhi University, Kerala, and has carried out extensive research in her subject area of Economics as well as in Education. Her research works on CCE, No detention policy, values & Global Citizenship Education (GCED), Systemic Reforms in school education, etc. have been published in journals/ books of international/national reputation and have been cited in UNICEF reports and elsewhere. At the doctoral level, four research scholars have been awarded PhD degrees under her guidance and supervision.

Sylvie Schoch is the Founder of IP-International GmbH – Creative Corporate Training, Master's degree in applied linguistics, adjunct professor at Bologna University, Italy, from 1998 to 2008, extensive experience as a corporate trainer and coach in Italy, Germany, France, and Canada. Areas of expertise include human resources development and creative training programme design.

Dr. Uttara Singh is an Assistant Professor in the Department of Geography, CMP Degree College, University of Allahabad, Prayagraj, Uttar Pradesh, India. Her areas of interest cover urban studies, planning, social and human geography, land use and sustainable regional development, remote sensing and geospatial studies. Dr. Singh has contributed to and published many research papers in journals of national and international repute.

Willeke van Staalduinen is a political scientist and was a nurse in practice. Her career can best be characterized as policy making in combination with health care and active ageing. Since 2017, she has been the co-founder of the private limited company AFEdemy, Academy on age-friendly environments, based in Gouda, the Netherlands, that focuses on the implementation of smart, healthy, age-friendly environments. AFEdemy provides research, curriculum development, training and advice on policy making.

Karin Stiehr, born in 1955, is the co-founder of ISIS GmbH, a private social research institute, based in Frankfurt am Main. The focus of her work is to foster the social integration and participation of vulnerable population groups, such as older people, care home residents, unemployed youth and refugees.

Adrian Heng Tsai Tan, PhD, is an Adjunct Lecturer with Curtin University (Singapore) and a Management Consultant in private practice. He earned a PhD in Management from the University of Canberra. His research areas include sustainability development, ethics, social responsibility, future of work and the interplay of education and work.

Lourdes Villardón-Gallego, PhD, is Full Professor at the Department of Education at the Faculty of Psychology and Education, University of Deusto, Spain. She teaches Educational Research and Developmental Psychology. Dr. Villardón-Gallego is the lead researcher of the eDucaR team. Her research work focuses on competences-based learning and collaborative environments for learning.

Concetto Paolo Vinci, PhD, is a Full Professor of Economics at the University of Salerno. He is member of Association for Comparative Economic Studies (ACES), www.acesecon.org. He earned his first degree from the University of Naples Federico II. Dr. Vinci earned his PhD in Development Economics, University of Naples Federico II. His main research interests are efficiency wages and dual economies, economics of migration, working hours and capital operating time.

Fathima Zehba M. P. is an Architect-Urban Planner currently pursuing her PhD at NIT Calicut. She earned a PG Degree in Urban Planning from NITC and a B. Arch Degree from Kerala University. She was a former Project Associate at the Urban Management Center (Consultant to Ministry of Housing and Urban Affairs, GoI) and an intern at National Institute of Urban Affairs, New Delhi. Research interests include Urban Quality of Life and its Assessment, Sustainable Living and Living conditions of Urban Poor.

1 Is the Human Development Index Still Appropriate for the Assessment of Quality of Life?

Adrian Heng Tsai Tan
Curtin University

CONTENTS

1.1 INTRODUCTION

The Human Development Index (HDI) has been widely accepted as a measure of quality of life for people living in a country (Biggeri and Mauro 2018; Hickel 2020). Developed by the United Nations Development Programme (UNDP), the intent of the HDI was to assess the development of a country through the lens of human capability development and potential: an alternative to the reliance of economic growth in determining country development (Sagar and Najam 1998; UNDP 2010). The HDI, which has ubiquitously focused on three key dimensions, namely health and life expectancy, education attainment and standard of living, has served as a proxy for the determination of quality of life in a country (McGillivray 1991; UNDP 2010, 2019).

A few decades ago, particularly since the late 20th century when many countries, especially in Asia, were on the road to development through rapid industrialization, the HDI in its current form to assess country development made utilitarian sense to uplift countries above the poverty line. For example, countries and territories such as Singapore, Hong Kong SAR, China, South Korea and Malaysia have recorded rapid rise in their HDI between 1990 and 2018 (UNDP 2018). These countries had reaped the benefits of the HDI to develop national policies which advanced their education, healthcare and economic systems to enhance human capability and potential of its citizens.

While the HDI has been an effective tool that has helped uplift countries to achieve basic standards of living—thus enabling its people to chart living progression on the basis of improvement in basic eudaemonic human development factors—the UNDP's latest 2019 Human Development Report (UNDP 2019) acknowledges deficiencies in the existing HDI algorithm and cites rising new-generation societal inequalities amid general improvements in HDI around the world. The rise of new-generation societal inequalities is partly due to the emergence of a variety of global mega trends—such as globalization, political upheavals, widening income gaps, climate change, shifts in cultural norms and values, disruptive technologies and epidemics due to emergence of new diseases and viruses—which influence well-being and the quality of life. Many researchers (Dasgupta and Weale 1992; Osberg and Sharpe 2005) have also argued against the use of the HDI—which has Gross National Income (GNI) per capita as an element in the index making it skewed towards a socioeconomic sphere of life—as an adequate measure of quality of life and consequently social well-being. While the GNI per capita and standard of living may be high, the quality of life may be low (Berenger and Verdier-Chouchane 2007).

The HDI appears in need of review and update so that it remains a sustainable system of human development which is adaptable with emerging global mega trends of disruption. Hence, it is not mere coincidence that the UNDP has highlighted the need for a more ecological and inclusive version of human development: one that pushes the potential to realize the Sustainable Development Goals (SDGs) of the United Nations (UNDP 2019). Interestingly, the SDGs that have been developed and adopted by member states of the United Nations in 2015 could serve as a driver of change, providing support for a revised HDI. This is because the SDGs as an integrated whole emphasize a vision to achieve inclusive development by uplifting human development through a balancing of social, economic and environmental dimensions of development (Gupta and Vegelin 2016; Hák et al. 2016; UNDP 2019): a philosophy that strongly resembles the original intention of human development based on the human capability approach (Sen 1999; Gasper 2002).

The need for review and update of the HDI also arises because of a potentially paradoxical relationship between HDI and quality of life. While HDIs may be high in a country, its citizens may not be experiencing a desirable quality of life due to the emergence of the variety of global mega trends highlighted earlier. In addition, global disruptive situations—and to a certain extent globalization—have caused high social stresses resulting from increased pressures of competition among people within a limited economic pie, a widening income gap, and the emergence of a digital divide (UNDP 2019). Additionally, the COVID-19 pandemic of 2020 at the time of writing this chapter has demonstrated how disruption in the form of a global health crisis can degenerate the way of life of citizens in a country, impacting national as well as individual well-being (UNDP 2020). Government measures both at the national and at the international level to mitigate the spread of the virus through a populace have partly contributed to such degeneration in the way of life of citizens around the world (Restubog et al. 2020). For instance, the lockdowns implemented in many countries have promoted isolation and hindered—although for valid reasons—human interaction, socialization and mobility, which are examples of contributing factors to social well-being of humanity (Nicola et al. 2020). In view of the uncertain and volatile situations just highlighted, the concepts of human development and quality of life require revisiting.

This chapter addresses the appropriateness of the HDI as a measure of quality of life, underpinned by theoretical foundations of quality of life and human development. This chapter is organized as follows. Firstly, the concepts of quality of life and human development are reviewed and discussed. Secondly, academic and practical perspectives of the HDI are critically analysed in relations to the review of the two aforementioned concepts to ascertain whether the current form of the HDI continues to be functionally appropriate as an assessment of quality of life. The analysis draws on secondary information available based on existential discussions in the academic research literature as well as reports of international organizations such as the United Nations and Organisation for Economic Co-operation and Development (OECD). Finally, propositions and a conceptual model are developed and discussed to highlight promising opportunities and directions for future research and development of the HDI.

1.2 METHODOLOGY

With the aim of addressing whether the HDI remains as an appropriate measure of quality of life and to propose a direction for future research and development in the field, a narrative review (Cronin et al. 2008) was adopted as the approach for performing a review and synthesis of the literature to address the contentious and complex research question. Several approaches for conducting a literature review of research studies and reports are available (Snyder 2019); however, a narrative approach was selected as the methodology that is most suited to interpret and critique existential studies on quality of life, human development and the HDI towards an engagement in theory development (Baumeister and Leary 1997). This is because as Sniltsveit et al. (2012, 414) suggests, "narrative reviews have become increasingly systematic, their methods have diversified and the terms to describe them have proliferated". In addition, "the emphasis on thinking and interpretation in narrative review" (Greenhalgh et al. 2018, 3) supports the aim to develop authoritative arguments to shape future research and development on quality of life, human development and the HDI.

In this research, various literature published between 1978 and 2020 were reviewed. The time period selected is justifiable since there is no definitive period of time over which a review of literature should be conducted; in addition, the body of literature for review is dependent on relevant studies that address the subject of interest (Cronin et al. 2008). With the intent to develop fresh and contemporary insights for future research and development of the HDI based on the evolution of theory development and existential controversies (Baumeister and Leary 1997), the narrative review was performed by providing a historical account, interpretation and critique of the development of two key concepts—quality of life and human development—which are relevant to the study of the HDI. We also integrated our understanding of quality of life and human development to critically analyse the literature which argue for and against the HDI within the specified time period. Both EBSCOhost and Google Scholar were the databases used in the search for academic literature. *Quality of life, human development* and *human development index* were the keywords used to query the databases. In addition, we also reviewed the reports that are available on the websites of the UNDP and OECD.

1.3   QUALITY OF LIFE: WHAT IS IT?

The concept of quality of life is multi-faceted and elusive (Wish 1986; Berenger and Verdier-Chouchane 2007). The proverbs "one man's meat might be another's poison" and "beauty is in the eyes of the beholder" might possibly provide the analogical reasoning to reflect the different perspectives that might exist concerning the meaning of quality of life and how the construct should be measured. Several factors could influence how quality of life is perceived and experienced. Quality of life as a construct, which is relative, might be perspective-dependent and perhaps even situational (McGregor and Goldsmith 1998). There are two dominant perspectives of quality of life: economic and healthcare perspectives (Dasgupta and Weale 1992). Events and situational factors arising from external environments such as the eruption of the COVID-19 pandemic, *ceteris paribus*, might moderate the status quo perceptions of quality of life (Nicola et al. 2020). It is also a social construct that is comprehended and experienced differently by various levels of society, which comprise the individual, communities and nations (McGregor and Goldsmith 1998; Pedro et al. 2020). Interestingly, cultural influences on quality of life do also exist (Woodside et al. 2020).

In Table 1.1, we provide a sample of various definitions and perspectives of quality of life that exist in contemporary academic discourse. The definitions by various authors are ordered in chronological order to elicit trends, similarities and differences in perspectives. A commonality among these definitions is the consideration of well-being indicators that are economic, social, environmental, psychological, physical and demographic in nature (McGregor and Goldsmith 1998; Cravioto et al. 2020). Limited differences exist among the definitions; however, there appears to be a trend to include an emphasis on intangibles such as longevity, freedom of choice, sustainable development and overall well-being. Researchers have also emphasized the need to differentiate quality of life, standard of living and well-being (McGregor and Goldsmith 1998; Berenger and Verdier-Chouchane 2007), while it is also interesting to highlight that quality of life has also been linked with well-being in that the latter forms the former (Dasgupta and Weale 1992; Berenger and Verdier-Chouchane 2007). There are also occasions in academic discourse when researchers have also referred to quality of life and well-being interchangeably (Wish 1986; Osberg and Sharpe 2005).

TABLE 1.1

Definitions and Perspectives of Quality of Life

Definitions and Perspectives of Quality of Life	Author(s)
The quality of life as measured by the Physical Quality of Life Index (PQLI) considers factors such as life expectancy, infant mortality and literacy as essential Measures for quality of life to develop desirable social distribution.	Morris (1978)
Quality of life does not just include socioeconomic factors of well-being such as literacy, life expectancy at birth, child survivability and real national income per capita which are devoted to national income and security, but also includes the availability of and access to civil liberties and political rights.	Dasgupta and Weale (1992)
Quality of life "consists of among other things, hope for the future, land, shelter, income, employment opportunities, maternal and child health, and family and social welfare".	McGregor and Goldsmith (1998, 2)
According to the capabilities approach, quality of life is achieved with the basic capabilities or freedom a person enjoys to Decide and pursue possible livings by choice.	Sen (1985, 1999); Saito (2003); Robeyns (2003)
A good life is defined by longevity and education.	Anand and Sen (2000)
Quality and quantity of life which economic well-being depends on is measured in terms of "yearly income and the number of years over which the income is enjoyed".	Becker et al. (2005, 277)
Quality of life refers to the well-being or human outcomes that facilitate the achievement of human freedoms by "emphasizing the 'being and doings' of a population, their opportunities and non-opportunities". It is comprised of quality of health, quality of education and quality of environment.	Berenger and Verdier-Chouchane (2007, 1262)
Quality of life is comprised of sustainable development and well-being.	Distaso (2007)

1.4 ABOUT HUMAN DEVELOPMENT

In order to provide a critical discourse of the HDI, it is imperative to understand the concept of human development. In association with the human capability approach proposed by Amartya Sen (1999), the UNDP has provided a definition of human development in its 1990 report (UNDP 1990, 10) as:

> Human development is a process of enlarging people's choices. The most critical ones are to lead a long and healthy life, to be educated and to enjoy a decent standard of living. Additional choices include political freedom, guaranteed human rights and self-respect.

In the 2010 Human Development Report (UNDP 2010), in acknowledgement of the dynamic nature of the human development approach and the need to be more aligned with academic literature and practical developments on the ground, the UNDP (2010, 2) had proposed an enhanced definition as:

> Human development is the expansion of people's freedoms to live long, healthy and creative lives; to advance other goals they have reason to value; and to engage actively in shaping development equitably and sustainably on a shared planet. People are both the beneficiaries and the drivers of human development, as individuals and in groups.

Since the launch of the first Human Development report in 1990, the focus on human development by the UNDP has undergone a conceptual evolution. Table 1.2 summarizes how the focus on human development has evolved.

From the humble beginnings of the report in 1990 (UNDP 1990) when the interest on human development was predominantly utilitarian in nature with focus on measurements and financing for the progress of human development, the report had gradually evolved to identify associations of human development with socio-political and external environment-mediating factors such as globalization, gender, economic growth, political development, human rights and technology just to name a few. Since 2011 and moving forward into the most recent 2019 and beyond, the conversation on human development has evolved to highlight the importance of acknowledging the global challenges of sustainability and equity (UNDP 2011, 2019) with the urgent need to manage inequalities by balancing variables of economic gains, social development and environmental impact. Nonetheless, human well-being, which bears the notion of humanity and freedom, has remained at the heart of the conversation on human development, and is very much still aligned with Amartya Sen's capability approach towards human development (Sen 1999). It is therefore interesting to highlight that the 2019 UNDP report on Human Development (UNDP 2019) had suggested that the perceptions of human development may have successfully veered too much towards a utilitarian economic angle with disregard for the total well-being of humanity.

In addition to publications by the UNDP and OECD, academic discourse is also replete with advancing discussion on the conceptualization and measurement of human development. For instance, Ranis et al. (2006) had proposed 11 categories of human development beyond what is currently measured by the HDI as basic aspects of human welfare. The research proposed the need to move beyond the measures for health (in terms of life expectancy), education (in terms of years of schooling) and standard of living (in terms of GNI per capita), and to include dimensions such as "mental well-being, empowerment, political freedom, social relations, community well-being, inequalities, work conditions, leisure conditions, political security, economic security and environmental conditions" (Ranis et al. 2006, 328–29) for the achievement of human flourishing and a full life. Herrero et al. (2012) have highlighted the substantive changes in perceptions towards human development in the UNDP 2010 Human Development Report (UNDP 2010) and proposed the need for further refinements due to shortcomings even in the improvements made in 2010. Many researchers (Ranis et al. 2006; Seth 2009) had also confirmed the multidimensional nature of human development with proposals for the need for more robustness in the conceptualization of human development. For example, Seth (2009)

TABLE 1.2

Chronological Development of the Focus of UNDP Human Development Reports

Year	Focus
2020	People and planet: Towards Sustainable Human Development *(this report is still under development at the time of acceptance of this book chapter)*
2019	Beyond income, beyond averages, beyond today: Inequalities in human development in the 21st century
2016	Human development for everyone
2015	Work for human development
2014	Sustaining human progress: Reducing vulnerabilities building resilience
2013	The rise of the South: Human progress in a diverse world
2011	Sustainability and equity: A better future for all
2010	The real wealth of nations: Pathways to human development
2009	Overcoming barriers: Human mobility and development
2007/2008	Fighting climate change: Human solidarity in a divided world
2006	Beyond scarcity: Power, poverty and the global water crisis
2005	International cooperation at a crossroads: Aid, trade and security in an unequal world
2004	Cultural liberty in today's diverse world
2003	Millennium development goals: A compact among nations to end human poverty
2002	Deepening democracy in a fragmented world
2001	Making new technologies work for human development
2000	Human rights and human development
1999	Globalization with a human face
1998	Consumption for human development
1997	Human development to eradicate poverty
1996	Economic growth and human development
1995	Gender and human development
1994	New dimensions of human security
1993	People's participation
1992	Global dimensions of human development
1991	Financing human development
1990	Concept and measurement of human development

Source: UNDP (Available from http://www.hdr.undp.org/en/global-reports).

highlighted the need for human development to be sensitive to forms of inequality. Neumayer (2012) argues for the importance of linkage of sustainability with human development, which should not be at the expense of environmental degradation.

1.5 IS THE HDI STILL RELEVANT?

The desire by policymakers to achieve equitable and holistic country development brings forth the question of whether the HDI is still relevant and meaningful from a national development perspective. Globally, particularly in the developed countries, rising inequalities beyond just income levels have caused the populace of diverse

backgrounds to challenge the seemingly known and desirable progress made in alleviating poverty and the achievement of basic standards of living (UNDP 2019). Highly
developed and prosperous territories or countries such as Hong Kong, Singapore and
United States of America have experienced their share of unhappiness among citizens who are possibly disconcerted with a predominantly economic-centric stance
of national development: a socio-psychological phenomenon possibly informing the
negative changes in levels of happiness or subjective well-being according to the
World Happiness Report 2020 (Helliwell et al. 2020). Citizens of such developed
societies are increasingly questioning the status quo perceptions about the achievement of economic progress and development at all cost; instead, there is movement
among the citizens towards a preference for a fairer and just society, and advocacy for
climate [Social and global justice (OECD 2019)]. However, while waves of government and policy disapprovals sweep across countries of varying stages of national
development—potentially highlighting issues with human development and the need
to address the contemporary relevance of the HDI (UNDP 2019; Repucci 2020)—it
might first be appropriate to address what the authentic intent of the HDI is.

Designed as a balanced approach for the measure of a country's development,
the HDI has been a prominent alternative to the gross domestic product (GDP),
which merely functioned as a perennial unidimensional measure of national
development (Sagar and Najam 1998). In recognition of the importance of enlarg-
ing choices of people in leading a long and healthy life, acquiring knowledge
through education and having access to resources to achieve a decent standard of
living, the HDI was initially developed to measure human development of a country along the lines of these considerations for a quality of life (UNDP 1990; Sagar
and Najam 1998). The index was intended as a mechanism for comparing progress
in human development by moving the debate on development beyond economic
measures (Morse 2003b). Presently, the motion could be considered successful as
many countries are now measured beyond GDP per capita in the UNDP Human
Development Reports.

Even though the HDI is a well-known yardstick of well-being (Klugman et al.
2011), it has remained a controversial indicator of human development since its inception in 1990 with arguments for its flaws and redundancies (McGillivray 1991; Cahill
2005). While the HDI is a welcomed alternative for the GDP as a measure of development, it has retained an economic measure in the form of the GNI per capita as an
indicator for the standard of living: a state that implies only a partial move beyond
solely economic measures of country development. It is widely recognized that economic measures of GDP, GNI and per-capita income are inadequate measures of
quality of life and well-being (Dasgupta and Weale 1992; Berenger and Verdier-
Chouchane 2007). According to Stiglitz et al. (2018), income figures may serve as
powerful economic indicators, but do not necessarily provide a holistic view about
the health of countries and societies, much less everything we need to know about
economic performance. While income measures do reflect basic human capability
to pursue a desired quality of life in terms of a decent standard of living—a concept
that was advocated by Amartya Sen's capability approach to a quality of life (Sen
1999) which originally informed the design of the HDI—income figures computed

as averages do not necessarily present a realistic representation of the income level of every citizen in a country (UNDP 2019).

The controversies associated with the computation of the HDI were formally recognized and addressed in UNDP research papers by Klugman et al. (2011) and Kovacevic (2011). There have also been developments in the interpretation of human development with increasing calls for a move away from the HDI in the country development debate (Morse, 2003b; Ranis et al. 2006; Herrero et al. 2012). Several researchers (Hickel 2020; Morse, 2003b; Kalimeris et al. 2020) continue to argue for the re-evaluation of the HDI as an appropriate assessment of quality of life and welfare. There is increasing advocacy for assessment of country development to be more inclusive of measures beyond economics and to embrace those of broader welfare for better assessment of well-being (Kalimeris et al. 2020).

In view of the recognition that human development encompasses more than just economics, the controversies and arguments against the current methodology of measuring HDI—which does not possibly reflect the original intent of Amartya Sen's capability approach to human development as advocate of freedom and agency of the actor—are intense (Sen 1985). Critics of HDI stress that "its indicators are too few and too arbitrarily chosen and that its definition is still inadequate and does not allow the capability approach to work" (Berenger and Verdier-Chouchane 2007, 1259). Even if proponents of the current HDI argue for the relevance of its three basic elements—health, education and (material) well-being—the appropriate variables would still need to be identified to meet the objectives of these key elements of human welfare (Herrero et al. 2012). In general, questions arise as to whether the component indicators, as listed in Table 1.3, are appropriate measures for the key dimensions of the current form of the HDI (UNDP 2010, 2019).

With an inference from Table 1.3, current indicators are clearly mere quantitative measurements of the associated dimensions of the HDI. While these were developed with good intentions, achievements in terms of good numbers may not necessarily suggest better health, education and well-being of humankind. We offer three reasons in the discussion that follows.

Firstly, a greater life expectancy does not necessarily represent that a person is living healthy years throughout the life expectancy particularly towards the end of life (Robine et al. 2009). Secondly, the number of years invested in education, particularly tertiary education, may not necessarily reflect the existence of quality in education: one that develops a path for a person to be able to compete and navigate in an increasingly challenging employment market characterized with limited opportunities for

TABLE 1.3

Current Dimensions and Indicators of the HDI

Dimension	Current Indicator
Health	Life expectancy at birth
Education	Expected years of schooling
	Mean years of schooling
Well-being	Gross GNI per capita

career advancement and achievement of aspirations due to rapidly changing nature of work caused by the proliferation of disruptive technologies (UNDP 2019).

Lastly, well-being is a concept that may not necessarily be best measured by income since it only refers to material well-being in terms of being able to afford a certain standard of living (Berenger and Verdier-Chouchane 2007). As discussed in the earlier section on quality of life, well-being is a multidimensional construct that includes aspects beyond material wealth (McGregor and Goldsmith 1998; Berenger and Verdier-Chouchane 2007; Cravioto et al. 2020). As a matter of fact, well-being is possibly comprised of two major components: quality of life and standard of living, as suggested by Berenger and Verdier-Chouchane (2007). However, both quality of life and standard of living have often been confused to mean the same; while the former refers more to the derivation of human outcomes, the latter refers more to material aspects of freedom with strong association with economic welfare and consumption (Sen 1984; Deutsch et al. 2003). Income measures have also generally been taken as a convenient indicator of well-being without acknowledging the shortcomings of such measures (Steckel 1995). An additional argument against the use of income measures for the HDI is the contradiction presented by growing economic and income inequality among the entire spectrum of countries of various levels of income, i.e. high, medium and low income, especially among the OECD countries, despite realizing improvements in HDI rankings (OECD 2008; UNDP 2013).

Despite updates since its inception in 1990, a critical review of the HDI by Sagar and Najam (1998) suggests that the HDI still "fails to include ecological considerations and (..) to capture the essence of the world it seeks to portray" (249)—an observation that remains relevant today. Hence, a critical need arises for a review of the relevancy of the HDI. Arguments against the HDI appear to propose either a review of the measurable indicators of existential key dimensions of the HDI or a total reinvention of the current form.

1.6 OPPORTUNITIES FOR FUTURE RESEARCH AND DEVELOPMENT OF THE HDI

With inference from the Human Development Report 2019 (UNDP 2019)—despite the narrowing gap in basic living standards made evident with the increasing numbers of people around the world escaping poverty, hunger and disease—the journey into the 2020s is marked with rising inequalities. The situation has resulted in the creation of a "new poor" among citizens of countries, including those of advanced economic status. Moreover, the COVID-19 pandemic at the time of writing of this chapter adds to the doldrums of the challenging situation that individuals need to cope with. In view of the arguments proposing for a review of the relevancy of the HDI for the assessment of human development and quality of life, this section proposes an agenda for future research and development as a potential contribution to a much-needed review. Based on the arguments presented in the earlier section of this chapter, it is proposed that the basis for future research and development of the HDI should encompass social inclusivity, ecological considerations, as well as resolve the contradiction between achieving a desirable HDI and rising income inequalities

(OECD 2008; UNDP 2019). Three propositions are formulated to shape the direction for future research and development of the HDI.

Proposition 1: A Future HDI Should Portray a Clear Distinction between Standard of Living and Quality of Life

It is proposed that a future HDI portrays a clear distinction between standard of living and quality of life. Based on the study by Berenger and Verdier-Chouchane (2007), it might appear that the current HDI is more of a measure of standard of living than quality of life. Hence, there is a need to review and re-incorporate the basis of Sen's human capability approach (Sen 1984, 1985, 1999) in the design of the HDI, and to emphasize more on outcomes of life and human development rather than material measures.

Proposition 2: A Review and Reinvention of the HDI Needs to Move Beyond Economics to Include the Concept of Human Flourishing

In view of the need to move beyond economics in the HDI, we propose the consideration of the concept of human flourishing—which is comprised of the suggested 11 elements explained in the earlier section on human development—as proposed by Ranis et al. (2006) in a revised HDI model. These elements, which signal a reinvigoration of the original intent of Sen's capability approach (Sen 1985, 1999) towards human development, are also highlighted in Table 1.4. In addition, as gender inequality continues to contribute to inequalities in the world (UNDP 2019), we recommend that the consideration of human flourishing does not neglect gender biases that exist and includes disaggregate information to study differential impacts on women. This

---

TABLE 1.4

Additional Elements for Human Flourishing for the HDI

- Mental well-being
- Empowerment
- Political freedom
- Social relations
- Community well-being
- Inequalities
- Work conditions
- Leisure conditions
- Political security
- Economic security
- Environmental conditions

Source: Derived from Ranis et al. (2006).

---

is so that any revised system of human development is inclusive and not one that is designed for men as the human default (Perez 2019).

Proposition 3: A Reinvented HDI Should Include Ecological and Sustainable Development Considerations

In view of proposed directions to include ecological considerations in the measurement of human development and quality of life, it is recommended that a reinvention of the HDI considers sustainable development by means of a sustainability index: an argument which several sustainable development researchers (Neumayer 2001; Morse 2003a; Distaso 2007; Moran et al. 2008; Biggeri and Mauro 2018; Hickel 2020) have persistently proposed since the inception of the HDI in 1990. Sustainable development through the provision of an ecological footprint (Moran et al 2008; Jeremic et al. 2011) is one of the "most significant indicator measuring a country's welfare" (Jeremic et al. 2011, 63), and hence quality of life and well-being. Sustainable development represents an essential commitment to "advancing human well-being, with the added constraint that this development needs to take place within the ecological limits of the biosphere" (Moran et al. 2008, 470). While a majority of ideas for a sustainable development element within the HDI appear to emphasize an environmentally conscious stance, there is another group of non-environment sustainability factors that require consideration. These factors relate to social elements within the realm of freedom such as political rights and civil liberties (Biggeri and Mauro 2018), and even considerations to assist low-income countries to sustain human development gains (Neumayer 2001). Several options of sustainability index have been proposed (Distaso 2007; Biggeri and Mauro 2018; Hickel 2020), and research for consolidation of these proposals will be necessary for a reinvention of the HDI. It is also suggested that the inclusion of sustainable development factors will substantially position a reinvented HDI closer to the United Nation's 2030 agenda for sustainable development by means of enabling the achievement of the SDGs (UNDP 2019).

In view of the proposed directions towards a review and reinvention of the HDI that was just discussed, a conceptual model to guide the development of a new HDI is proposed in Figure 1.1. The conceptual model emphasizes Sen's capability approach (Sen 1985, 1999; Diastoso 2007) as the basic driver for the attainment of quality of life. The model illustrates an outcome-based process approach starting with the application of Sen's theory of well-being (Diastoso 2007) to derive economic, and social and environmental outcomes towards the achievement of sustainable human development, which serves as an antecedent of quality of life.

1.7   CONCLUSIONS

The purpose of this chapter was to provide a critical discussion on the appropriateness of the HDI as a barometer for the quality of life in countries. Two key concepts, namely quality of life and human development, were critically reviewed. The review highlighted several definitions and perspectives in relation to both concepts. While

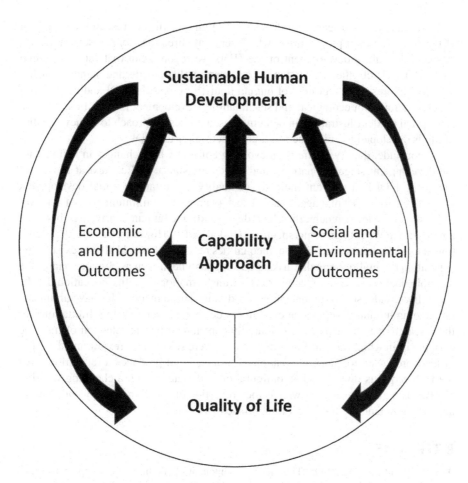

FIGURE 1.1 Outcome-based approach to research and development of a reinvented HDI.

there were differences in perspectives, the chronological development of each concept uncovered similarities. Quality of life tended to be interchangeably referred to as well-being or standard of living although the academic literature is clear about the differences. There also appears to be a trend to include intangibles in the consideration of quality of life. The discourse on human development while rooted in Sen's capability approach (Sen 1985, 1999) has evolved to encompass more non-economic considerations such as freedom and sustainable development.

In integrating the discourse presented for both quality of life and human development, this chapter provided an assessment of the appropriateness of the HDI and noted that it remains a controversial indicator of human development. While Sen's capability approach (Sen 1985, 1999) was initially used as the basis for the development of the HDI, there is motivation for greater inclusion of sustainable development measures into the HDI since the indicators that were originally developed and are presently used may not necessarily provide an accurate assessment of both human development outcomes and the evolving nature of human development.

In summary, our research produces suggestions for further research in the form of the three research propositions, which serve as directions for a research agenda to shape the future development of the HDI. We recommend that future research on the HDI (1) clarifies the dynamics between quality of life and standard of living, (2) explores the integration of human flourishing into the HDI and (3) includes the consideration of ecological and sustainability development factors. In addition, a conceptual model illustrating an outcome-based process approach to achieve quality of life is developed to guide the research direction for the HDI.

To conclude, quality of life is an evolving concept which changes in nature with the development stages of nations, and the circumstance and context at a point in time. The HDI is a relevant indicator of quality of life at the initial development stages of a nation as it manages to uplift society above the minimum floor of achievement in human development, thereby reducing extreme human deprivation, but from a predominantly economic perspective (UNDP 2019). However, beyond the reduction of extreme human deprivation, inequalities remain and would always exist. New-generation inequalities are also arising due to phenomena such as climate change, the technological divide due to technological transformations, quality education divide, capability divide, social divide and power divide. These inequalities have an impact on ongoing human development even in the developed world. While limitations for the research in this chapter are potentially contributed by the selection of the time period within which various kinds of literature were selected for review and the types of keywords used to query the databases, the research produces a recognition that the HDI requires review and reconfiguration. Adaptation with a changing world to resolve rising inequalities, as well as the promotion of social inclusivity and sustainable development, are the tenets for a reinvented HDI.

REFERENCES

Anand, S. and Sen, A. 2000. "The income component of the human development index." *Journal of Human Development* 1(1): 83–106.

Baumeister, R.F. and Leary, M.R. 1997. "Writing narrative literature reviews." *Review of General Psychology* 1(2): 311–320.

Becker, G.S., Philipson, T.J. and Soares, R.R. 2005. "The quantity and quality of life and the evolution of world inequality." *American Economic Review* 95(1): 277–91.

Berenger, V. and Verdier-Chouchane, A. 2007. "Multidimensional measures of well-being: Standard of living and quality of life across countries." *World Development* 35(7): 1259–76.

Biggeri, M. and Mauro, V. 2018. "Towards a more 'sustainable' human development index: Integrating the environment and freedom." *Ecological Indicators* 91: 220–231.

Cahill, M.B. 2005. "Is the human development index redundant?" *Eastern Economic Journal* 31(1): 1–5.

Cravioto, J., Ohgaki, H., Che, H.S., Tan, C.K., Kobayashi, S., Toe, H., Long, B., Rahim, N.A. and Farzeneh, H. 2020. "The effects of rural electrification on quality of life: A Southeast Asian perspective." *Energies* 13(10): 2410.

Cronin, P., Ryan, F. and Coughlan, M. 2008. "Undertaking a literature review: A step-by-step approach." *British Journal of Nursing* 17(1): 38–43.

Dasgupta, P. and Weale, M. 1992. "On measuring the quality of life." *World Development* 20(1): 119–131.

Distaso, A. 2007. "Well-being and/or quality of life in EU countries through a multidimensional index of sustainability." *Ecological Economics* 64(1): 163–180.

Deutsch, J., Ramos, J. and Silber, J. 2003. "Power and inequality of standard of living and quality of life in Great Britain." In *Advances in Quality of Life Theory and Research*, edited by E. Diener and D. Rahtz, 99–128. Dordrecht: Springer.

Gasper, D. 2002. "Is Sen's capability approach an adequate basis for considering human development?" *Review of Political Economy* 14(4): 435–461.

Greenhalgh, T., Thorne, S. and Malterud, K. 2018. "Time to challenge the spurious hierarchy of systematic over narrative reviews?" *European Journal of Clinical Investigation* 48(6): 1–6.

Gupta, J. and Vegelin, C. 2016. "Sustainable development goals and inclusive development." *International Environmental Agreements: Politics, Law and Economics* 16(3): 433–448.

Hák, T., Janoušková, S. and Moldan B. 2016. "Sustainable development goals: A need for relevant indicators." *Ecological Indicators* 60(2016): 565–573.

Helliwell, J.F., Lanyard, R., Sachs, J. and De Neve, J. eds. 2020. *World Happiness Report 2020*. New York: Sustainability Development Solutions Network.

Herrero, C., Martinez, R. and Villar, A. 2012. "A newer human development index." *Journal of Human Development and Capabilities* 13(2): 247–268.

Hickel, J. 2020. "The sustainable development index: Measuring the ecological efficiency human development in the Anthropocene." *Ecological Economics* 167: 106331.

Jeremic, V., Isljamovic, S., Petrovic, N., Radojicic, Z., Markovic, A. and Bulajic, M. 2011. "Human development index and sustainability: What's the correlation?" *Metalurgia International* 16(7): 63–67.

Kalimeris, P., Bithas, K., Richardson, C. and Nijkamp, P. 2020. "Hidden linkages between resources and economy: A "Beyond-GDP" approach using alternative welfare indicators." *Ecological Economics* 169: 106508.

Klugman, J., Rodrigeuz, F. and Choi, H.J. 2011. "The HDI 2020: New controversies, old critiques." *Human Development Research Paper* 1: 1–45.

Kovacevic, M. 2011. "Review of HDI critiques and potential improvements." *Human Development Research Paper* 33: 1–44.

McGillivray, M. 1991. "The human development index: Yet another redundant composite development indicator?" *World Development* 19(10): 1461–1668.

McGregor, S.L.T. and Goldsmith, E.B. 1998. "Expanding our understanding of quality of life, standard of living, and well-being." *Journal of Family and Consumer Services* 90(2): 2–6.

Moran, D.D., Wackernagel, M., Kitzes, J.A., Goldfinger, S.H. and Boutaud, A. 2008. "Measuring sustainable development – Nation by nation" *Ecological Economics* 64(3): 470–474.

Morris, M.D. 1978. "A physical quality of life index." *Urban Ecology* 3(3): 225–40.

Morse, S. 2003a. "Green the United Nations' human development index?" *Sustainable Development* 11(4): 183–198.

Morse, S. 2003b. "For better or for worse, till the human development do us part?" *Ecological Economics* 45(2): 281–296.

Neumayer, E. 2001. "The human development index and sustainability—a constructive proposal" *Ecological Economics* 39(1): 101–114.

Neumayer, E. 2012. "Human development and sustainability." *Journal of Human Development and Sustainability* 13(4): 561–579.

Nicola, M., Alsafi, Z., Sohrabi, C., Kerwan, A., Al-Jabir, A., Losifidis, C., Agha, M. and Agha, R. 2020. "The socio-economic implications of the coronavirus pandemic (COVID-19): A review." *International Journal of Surgery* 78: 185–193.

OECD (Organisation for Economic Cooperation and Development). 2008. *Growing Unequal? Income Distribution and Poverty in OECD Countries.* Paris: OECD Publishing. https://doi.org/10.1787/9789264044197-en.

OECD (Organisation for Economic Cooperation and Development). 2019. *Development Co-Operation Report 2019: A Fairer, Greener, Safer Tomorrow.* Paris: OECD Publishing. https://doi.org/10.1787/9a58c83f-en.

Osberg, L. and Sharpe, A. 2005. "How should we measure the "economic" aspects of well-being?" *Review of Income and Wealth* 51(2): 311–36.

Pedro, E.D.M., Leitao, J. and Alves, H. 2020. "Bridging intellectual capital, sustainable development and quality of life in higher education institutions." *Sustainability* 12(2): 479.

Perez, J.C. 2019. *Invisible Women: Exposing Data Bias in a World Designed for Men.* London: Penguine Random House.

Ranis, G., Stewart, F. and Samman E. 2006. "Human development: Beyond the human development index." *Journal of Human Development* 7(3): 323–358.

Repucci, S. 2020. "A leaderless struggle for democracy." In *Freedom in the World 2020*, edited by N. Buyon, I. Linzer, T. Roylance and A. Slipowitz, 1–15. Washington, DC: Freedom House.

Restubog, S., Ocampo, A. and Wang, L. 2020. "Taking control amidst the chaos: Emotion regulation during COVID-19 pandemic." *Journal of Vocational Behaviour* 119: 103440.

Robeyns, I. 2003. "Sen's capability approach and gender inequality: Selecting relevant capabilities." *Feminist Economics* 9(2–3): 61–92.

Robine, J., Saito, Y. and Jagger, C. 2009. "The relationship between longevity and healthy life expectancy." *Quality in Ageing and Older Adults* 10(2): 5–14.

Sagar, A.D. and Najam, A. 1998. "The human development index: A critical review." *Ecological Economics* 25(3): 249–264.

Saito, M. 2003. "Amartya Sen's capability approach to education: A critical exploration." *Journal of Philosophy of Education* 37(1): 17–33.

Sen, A. 1984. "The living standard." *Oxford Economic Papers* 36: 74–90.

Sen, A. 1985. "Well-being, agency and freedom: The Dewey lectures 1984." *The Journal of Philosophy* 82(4): 169–221.

Sen, A. 1999. "Development as a freedom." In *The Globalization and Development Reader: Perspectives on Development and Global Change*, edited by J. Timmons Roberts, Amy Bellone Hite and Nitsan Chorev, 525–48. Oxford: John Wiley & Sons.

Seth, S. 2009. "Inequality, interactions, and human development." *Journal of Human Development and Capabilities* 10(3): 375–396.

Sniltsveit, B., Oliver, S. and Vojtkiva, M. 2012. "Narrative approaches to systematic review and synthesis of evidence for international development policy and practice." *Journal of Development Effectiveness* 4(3): 409–429.

Snyder, H. 2019. "Literature review as a research methodology: An overview and guidelines." *Journal of Business Research* 104: 333–339.

Steckel, R.H. 1995. "Stature and the standard of living." *Journal of Economic Literature* 33(4): 1903–1940.

Stiglitz, J., Fitoussi, J. and Durand M. 2018. *Beyond GDP: Measuring What Counts for Economic and Social Performance.* Paris: OECD Publishing.

UNDP (United Nations Development Programme). 1990. *Human Development Report 1990: Concepts and Measurement of Human Development.* New York: Oxford University Press http://www.hdr.undp.org/en/reports/global/hdr1990.

UNDP (United Nations Development Programme). 2010. *Human Development Report 2010: The Real Wealth of Nations: Pathways to Human Development.* New York: Palgrave Macmillan. http://www.hdr.undp.org/en/content/human-development-report-2010.

UNDP (United Nations Development Programme). 2011. *Human Development Report 2011: Sustainability and Equity - A Better Future for All Human Development Reports.* New York: Palgrave Macmillan. http://www.hdr.undp.org/en/content/human-development-report-2011.

UNDP (United Nations Development Programme). 2013. *Humanity Divided: Confronting Inequality in Developing Countries, United Nations Human Development Programme.* New York: UNDP Bureau for Development Policy.

UNDP (United Nations Development Programme). 2018. *Human Development Data (1990–2018).* United Nations Human Development Programme. http://hdr.undp.org/en/data.

UNDP (United Nations Development Programme). 2019. *Human Development Report 2019: Beyond Income, Beyond Averages, Beyond Today: Inequalities in Human Development in the 21st Century.* New York: AGS. http://www.hdr.undp.org/en/content/human-development-report-2019.

UNDP (United Nations Development Programme). 2020. *Coronovirus vs Inequality.* United Nations Human Development Programme. https://feature.undp.org/coronavirus-vs-inequality/.

Wish, W.B. 1986. "Are we really measuring the quality of life? Well-being has subjective dimensions, as well as objectives ones." *American Journal of Economics and Sociology* 45(1): 93–99.

Woodside, A.G., Megehee, C.M., Isaksson, L. and Ferguson, G. 2020. "Consequences of national cultures and motivations on entrepreneurship, innovation, ethical behaviour, and quality of life." *Journal of Business and Industrial Marketing* 35(1): 40–60.

2 Workers' Well-Being and Workplace Environment
Crossed Case Studies of Workplaces in Italy, Greece and Portugal

João Leitão
University of Beira Interior
University of Lisbon

Dina Pereira
University of Beira Interior
University of Lisbon

Ângela Gonçalves
University of Beira Interior

CONTENTS

2.1 INTRODUCTION

Well-being has moved into the centre stage, in recent years, by both scholars and practitioners concerned with the development of a meaningful and sustainable society and
that have become increasingly dissatisfied with purely financial measures of progress.

There is a convergent understanding in the reference literature (e.g. Ryan and
Deci 2001; Sirgy et al. 2001) that well-being is more than the absence of negative or
ill health states and is usually defined as the presence of positive feelings and functioning. The study of well-being at work is multidisciplinary, including economics,
management, industrial and organizational psychology, anthropology and sociology
(e.g. Day and Randell 2015; Bakker 2015; Cooper and Leiter 2017; Leitão et al. 2019).

Work-related well-being is a component of subjective well-being, which is positively influenced by physiological components of work, such as satisfaction, morale,
engagement and involvement, with psychosomatic and health issues, for example,
mental disorders (e.g. depression and burnout), tension at work, having a negative
impact on work-related well-being.

This chapter is innovative in the sense that it analyses how (and to what extent) the
environmental and microclimatic conditions, as well as the organization of employees' daily routines and tasks, impact on workers' well-being, by making use of a
Heart Rate Variability (HRV) measure. In this line of action, it is compared the set
of well-being conditions of the staff working in an innovative work environment,
with those in a standard workplace, in three different countries and workplaces, one
in Greece, one in Portugal and two in Italy.

The remainder of this chapter is structured as follows. Firstly, a selected set of
theoretical underpinnings on well-being is presented. Second, the research methodology is presented. Next, the presentation and discussion of results is made available,
followed by the conclusions, limitations and guidelines for future research.

2.2 THEORETICAL UNDERPINNINGS ON WELL-BEING

2.2.1 DEFINING AND CRAFTING ENVIRONMENTAL WORKPLACE CONDITIONS

Defining workplace well-being is not a simple task, as it has a multidimensional nature
(Dodge et al. 2012). Workplace well-being embraces physical health and comfort,
mental health, positivism and positive attitudes to work (Cooper and Leiter 2017).

The topic on workplace environmental conditions is at the core of ongoing debate,
as those conditions that affect workers' health and well-being, influencing absenteeism
and productivity (Im et al. 2018; Leitão et al. 2019). Several risks at the workplace may
negatively influence workers' well-being, for example, towards indoor environmental
quality (e.g. carcinogens, air pollution, noise, radiation, etc.) (WHO 2012; Ji et al. 2018).

The set of guiding mitigation measures is regulated by several international entities, e.g. the Occupational Safety and Health Administration (OSHA) and World
Health Organization (WHO), which provided guidelines, programs and manuals
to be used by organizations and also legislation to regulate companies' safety conditions, employee training and workplace conditions (Aryal, Parish, and Rohlman
2019; WHO 2016; IAEA and ILO 2018).

A set of factors may affect workers' well-being, namely environmental conditions, motivational factors, psychosocial and economic variables (Aziri and Irfan 2011; Bakotić and Tomislav 2013; Emrah et al. 2014; Hanaysha and Tahir 2016; Maghsoodi et al. 2018; Raziq and Maulabakhsh 2015; Leitão et al. 2019).

For workers to experience a high level of well-being, a set of positive organizational attitudes must be secured (Keeman et al. 2017), including higher work performance (Lyubomirsky, King, and Diener 2005), low turnover intentions, low actual turnover (Boehm and Lyubomirsky 2008), greater effort and thought put into work, less absenteeism and fewer work-related injuries (Keyes and Grzywacz 2005). As pointed out by several scholars, well-being is affected by work and issues related to organizational performance. Therefore, it is of the utmost importance that an organization supports and promotes well-being at work (Dewe and Cooper 2012; Hone et al. 2015).

Traditionally, organizations have centred their strategic attitudes linked with employee well-being on stress reduction more than on promoting employee well-being (Hone et al. 2015), as this is more correlated with decreased productivity and reduced profit (Ford et al. 2011), making organizations more prone to implement stress reduction interventions (Kelloway and Day 2005). Stress and well-being are linked constructs. In fact, investing in promoting workers' well-being may decrease organizational stress (Hone et al. 2015).

A definition of well-being comprehends two main perspectives, a hedonic perspective, describing well-being as happiness, and putting the focus on life satisfaction, the presence of positive mood, and the absence of negative mood (Ryan and Deci 2001) and an eudemonic perspective, where well-being is related to self-actualization, arguing that true happiness is found in expressing virtue (Dewe and Cooper 2012). Well-being can be the combination of feeling good (hedonism) and functioning well (eudemonism) (Aked et al. 2009).

Prior research indicates that workers feeling higher levels of well-being tend to make more effort at work (Keyes and Grzywacz 2005; Day and Randell 2014) than those revealing poor psychological health, such as depression, anxiety, fatigue and lack of motivation to engage in positive behaviours at work. Workers' well-being is associated with diverse positive dynamics at the workplace, namely team cohesion and engagement (Bakker 2015).

Monitoring environmental parameters inside the workplace is of extreme importance to promote health and wellness conditions. Microclimate refers to the set of physical and climatic parameters in the workplace (such as temperature, relative humidity or air speed). For better illustrating this statement, the spread of microbiological contaminants in a closed space is influenced by temperatures above 26°C and relative humidity above 65%. The thermal comfort zone in sedentary working conditions and seasonal clothing is around 22°C, relative humidity between 40% and 60% (summer), and around 19.5°C, with relative humidity of 40%–60% (winter) (Nanni, Benetti, and Mazzini 2017).

The possibility of employees having greater local control over their environmental conditions has a positive effect on their performance, motivation, well-being and overall productivity (Burge et al. 1987; Leitão et al. 2019).

2.2.2 DESIGNING AND ORGANIZING HEALTHY WORKING CONDITIONS

The Job Characteristics Model, proposed by Hackman and Oldham (1976), outlines a set of five core job characteristics, namely skill variety, task identity, task significance, autonomy and feedback, all impacting favourably on the organizational performance, by focusing on workers' feelings, motivation and job satisfaction.

Poorly designed layouts can result in low productivity, operator fatigue and increased likelihood of mistakes occurring (Marmaras and Nathanael 2006). Interestingly, little efforts coming from organizational economics and management research have been devoted to examine the effects of the physical work environment itself on employees (Carlopio 1996; Cohen 2007).

Workers' management is also a core aspect for their well-being, with an adequate supervisory model affecting employee engagement, as the supervisor is viewed as the most trusted source of information in the workplace (Kahneman et al. 2004).

Workplace design will affect to a great extent the postures that the worker will be able to adopt (Marmaras and Nathanael 2006). Consequently, building design needs to consider the parameters of occupants' well-being right at the beginning (Al horr et al. 2016). Architectural design has a direct impact on office lighting, and office lighting has a direct impact on well-being and productivity. Visual comfort is also very important for the well-being and productivity of buildings' occupants (Leech et al. 2002; Serghides, Chatzinikola, and Katafygiotou 2015).

Research on the physical characteristics of workplaces, for example, lighting, noise and air quality, shows that physical characteristics can influence employee health (Vischer 2007). Research has also been carried out on the effects of exposure to natural elements, such as greenery and sunlight, on physical and mental health and its importance in lowering stress and increasing psychological well-being (Berman, Jonides, and Kaplan 2008; Hartig et al. 2003; Holick 2004; Knight and Haslam 2010; Leather et al. 1998; Nieuwenhuis et al. 2014; Velarde, Fry, and Tveit 2007). Moreover, indoor environment quality has a major impact on occupant productivity and occupant behaviour (Al horr et al. 2016).

Work organization influences employees both physically and psychosocially. Physically, work systems affect the work environment, ergonomic factors and the workspace layout (Carayon and Smith 2000). Work organization may also affect some psychosocial aspects, including stress and strain, which can lead to employee burnout, accidents or under-performance (Carayon and Smith 2000; Edwards et al. 2009). Furthermore, the work organization system influences employees' health, well-being, behaviours, satisfaction and performance (Bakotić and Tomislav 2013; Gou 2019; Ji, Pons, and Pearse 2018; Kottwitz et al. 2018; Kwon, Lim, and Lee 2018; Lee and Brand 2005; Salonen et al. 2013), and ultimately represents a cost for organizations (Edwards et al. 2009).

2.3 RESEARCH METHODOLOGY

2.3.1 AIMS

This innovative study aims to reveal the associations between workers' well-being and their workplace conditions, providing new insights for possible future implementations in workplace organization. For accomplishing this, an innovative experiment was set up and conducted in order to compare the set of well-being conditions of the personnel working in an innovative workplace environment (using innovative solutions for the work environment, such as microclimatic intelligently controlled conditions, desk and space reservation for meetings, activity planning, among others), with others considered to be standard workplaces.

2.3.2 DESK RESEARCH

The innovative experiment was preceded by the development of a survey funded on different questionnaires previously used to carry out related surveys on health and well-being in the workplace, including the pioneering measure on quality of work life developed by Sirgy et al. (2001), and the set of analytical tools surveyed and empirically operationalized by Leitão et al. (2019).

The survey was conducted between April and July of 2018. Twelve partners from Italy, Bulgaria, Cyprus, Portugal, Greece and Spain took part interviewing employees. The sample covered 15 private companies and five public entities or large firms per partner, involving two employees per organization, totalling 443 questionnaires. It was not intended to interview neither the owners of companies nor general managers to avoid biases in the responses. A convenience sample procedure, under a random basis, was applied.

This survey made possible to identify several factors that are potential influencers of the desire of employees to contribute (or not) to the organizational productivity (Leitão et al. 2019). Furthermore, it raised unexplored factors related to the stress, and the physiological and psychosomatic condition of employees, as well as their linkages with the role played by environmental and health conditions at the workplace, in promoting well-being at the workplace as an organizational lever for increasing satisfaction, productivity and performance.

2.3.3 EXPERIMENT SET-UP

The experiment embraced monitoring activities, which were carried out for 2 weeks, in three European Countries (Italy, Greece and Portugal). These workers' activities comprised essentially of telephone calls, planning the working day, debriefing with employees, assigning tasks, taking decisions, listening to others' presentations, doing office-related tasks like writing on the computer, printing documents, etc. Perceived well-being was evaluated using thermo-technical parameters such as temperature, air quality, indoor airflow and lighting. The workplaces used in the experiment were the following:

- **Masterandskills' headquarters in Sapienza University of Rome**: Italy, which is a standard workplace since the building was built in the 1950s and is characterized by a standard layout (aisles and assigned rooms) and standard work organization.
- **eFM's headquarters in Rome**: Italy, where the workplace has been used since 2016 according to the most innovative solutions: no fixed desks for personnel; flexibility of desk occupancy according to the activities performed

by the personnel (working alone, in small groups, meeting, presentations); microclimatic conditions set according to the occupancy of each desk; environmental noise; planning of desk occupancy through specific apps available on the smartphone; green inside the headquarters useful to reduce the electromagnetic field generated by PCs.

- **Tsikrikoni's headquarters in Nea Kallikratia**: Greece, a Real Estate Agency located in Chalkidiki Peninsula, near Makedonia close to Thessaloniki International Airport. Since 1989, Tsikrikoni has provided some of the best properties on the Chalkidiki Peninsula, a popular tourist area in northern Greece, famous for its resorts and natural beauty. This geographically favourable position allows the agency to get in touch and collaborate with people from all over the world.
- **UBImedical's headquarters in Covilhã**: Portugal is a space of excellence to articulate the connection between the University and the business world. Created to expedite the transfer of knowledge in the search for new technologies, it allows companies to develop the research and laboratory tests necessary to, effectively, commercialize new products, generating benefit for the economy. UBImedical comprises two distinct areas: the laboratory area and the incubation area, providing the best conditions to develop projects and offering premium physical spaces at competitive prices.

The participants in the experiment were divided into small groups and located in the workplaces where the experimentation/internship took place. One of these groups (nine trainees) worked both at eFM's Headquarters in Rome and at Masterandskills' Headquarters and training place in Sapienza University of Rome (1 week at eFM and 1 week at Masterandskills) to allow direct comparison between the working conditions in these places. The same group of employees worked at eFM Headquarters and was relocated for a week at Masterandskills (performing the same activities), in order to record the differences in well-being due to the different workplace. In Greece and Portugal, the employees worked in one place only.

The protocol of tasks performed by the trainees was as follows:

- support employees working in the selected workplaces to monitor of well-being parameters during the working day;
- download data and information from the wearable to PCs in order to allow further data processing;
- define and implement small changes to the workplace layout and/or work organization, in order to improve employees' well-being;
- analyse day by day the changes in employees' well-being due to these small adjustments.

2.3.4 Sample Characterization

In Italy, nine EFM employees, four males and five females, were involved in the study. Their ages range from 27 to 60. The range can be subdivided in three categories:

- under 30 years: 22.5%;
- from 30 to 40 years: 55%;
- over 40 years: 22.5%.

In Greece, two employees were involved, one male and one female. Their ages range from 35 to 50.

In Portugal, there were also two employees involved, one male and one female. Their ages range from 29 to 39.

2.3.5 HRV Measurements

HRV is biometric data related to variation of the heart rate. It is important to understand HRV correctly in order to have correct interpretation of the data collected in the different workplaces. HRV does not refer to the heart rate itself; rather, it shows the variation between one heartbeat and the next, according to the time between them, in terms of milliseconds.

Scientific studies have shown that the time between heartbeats is not constant, more precisely that time is always variable, in terms of milliseconds (Stein et al. 1994; Bailon et al. 2007; Acharya et al. 2006). Everyone presents a natural variability in heart rate according to factors such as breathing rhythm, emotional states, anxiety, stress, anger, relaxation and so on.

HRV is a useful signal to understand the status of the autonomic nervous system (ANS) (Acharya et al. 2006). The sympathetic nervous system, when activated, produces a series of effects such as acceleration of the heartbeat, dilation of the bronchi, increase in arterial pressure, peripheral vasoconstriction, pupillary dilation and increased sweating.

To measure participants' HRV, specific equipment consisting of a strap – the HRV4Training from POLAR, which is fastened around the chest, collected the data. The strap consists both of a plastic part, which is slightly moistened, and of one transmitter, linked to the strap at two points. The strap is in contact with the skin, and the transmitter is able to record the variation of the heartbeat. These variations are transmitted to the smartphones where an app can visualize the data. The registrations were made every 30 minutes, and they were recorded on a sheet, together with the other information about the type of activity carried out, the personal feeling toward the environment, and the place where it occurred. To visualize the HRV value on the smartphone screen, the screen itself is pressed, and after about 2 minutes, the HRV value appears. Subsequently, a manual recording was required.

In Greece and Portugal, the tutor recorded employees' HRV on a sheet of paper and this was linked to the activities performed in the workplace and the climatic conditions. In Italy, this was done directly by the employees and the tutors supported them during data collection. The data collection was performed every 30 minutes, considering several parameters:

- HRV value;
- Main activities performed;
- Working conditions;
- Climatic and environmental conditions.

To measure HRV, every employee used a band and a smartphone app. The band used was the polar H7. An app was downloaded to the tutor's phone, in order to observe measuring of the HRV. The app in question was HRV4Training, and the results of the measurements were presented on the smartphone screen.

The HRV HRV4Training app uses rMSSD to compute the HRV4Training Recovery Points. rMSSD is computed as the square root of the mean squared differences of successive RR intervals. When computing rMSSD, we are looking at beat-to-beat differences; thus, the rMSSD feature is associated with short-term changes in the heart. Since parasympathetic activity works at a faster rate (e.g. < 1 second) than sympathetic activity, rMSSD is considered a solid measure of vagal tone and parasympathetic activity. rMSSD values are transformed to make the values more user-friendly. The result is an approximate number on a scale between 1 and 10, with higher values representing higher parasympathetic activity, lower stress and better recovery (in general).

HRV, in particular rMSSD or a transformation of rMSSD, such as HRV4Training's Recovery Points, are simply a way to capture parasympathetic activity, or in other words, the level of physiological stress.

In order to understand the correlation between the HRV value and the activities performed, it is important to describe the main activities of the sample of workers under study. Among others, four different types of activities were outlined: (1) working, (2) management, (3) presentation and (4) relationship with colleagues and clients.

Firstly, regarding working activities, a set of activities was identified, namely telephone calls, planning of the working day, conference video calls, writing/printing documents and working with a personnel computer (depending on the employee's role).

Secondly, for management activities, the following were indicated: assigning tasks to employees; controlling work performed by other people; debriefing with employees; coordinating the team; and decision-making about work-related activities.

Thirdly, related to presentation activity, actions to other people-oral speech and to other people-listening were listed.

Fourthly, the activity of relationship with colleagues and clients was reported through two main types of work: talking about work and talking about other subjects.

In addition to the main activities performed, the related work conditions were taken into account. Those conditions were working alone (to study), working alone (to create something new), working alone (administrative duties), working in a group, working under pressure and working without pressure.

2.4 RESULTS AND DISCUSSION

Table 2.1 provides information about the average HRV (for the 2 weeks) for each employee with respect to the location where the study was performed. As stated previously, in Italy the tests were performed in two different locations (eFM and MasterandSkills).

Figure 2.1 reveals that the difference in HRV values is not substantial, especially, for the employees in the two workplaces. The orange line indicates MasterandSkills, and the other indicates eFM headquarters. No chart is given for the employees in Portugal and Greece, because each trainee only followed employees in one specific location. The employees did not relocate, as happened in MasterandSkills and eFM.

TABLE 2.1
Average HRV Value for Each Employee

Employees	Locations	Average HRV	Employees	Location	Average HRV
Emp-1-G	Greece	7.7			
Emp-2-G	Greece	7.8			
Emp-1	Italy-eFM	7.5	emp-1	Italy- Masterandskills	7.1
Emp-3	Italy-eFM	6.3	emp-3	Italy- Masterandskills	0
Emp-4	Italy-eFM	6.5	emp-4	Italy- Masterandskills	6.5
Emp-5	Italy-eFM	7.5	emp-5	Italy- Masterandskills	7.3
Emp-6	Italy-eFM	6.7	emp-6	Italy- Masterandskills	6.6
Emp-7	Italy-eFM	6.5	emp-7	Italy- Masterandskills	0
Emp-8	Italy-eFM	6	emp-8	Italy- Masterandskills	0
Emp-10	Italy-eFM	6.4	emp-10	Italy- Masterandskills	6.3
Emp-1-P	Portugal	6.3			
Emp-2-P	Portugal	7.7			

Source: Own elaboration.

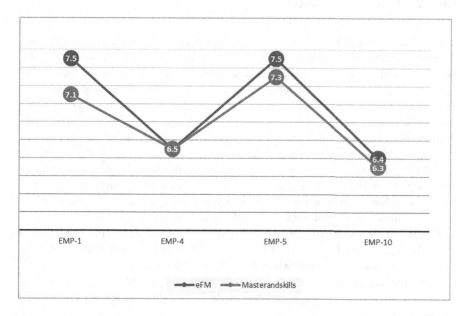

FIGURE 2.1   HRV average values at Masterandskills and eFM.

Figure 2.2, illustrating the trend line with respect to HRV values in all locations, shows that the employees with the lowest HRV are from Italy (emp-3, emp-4, emp-7, emp-8 and emp-10) and Portugal (emp-1-P). It is also observed that both the employees from Greece (emp-1-G and emp-2-G) have the highest HRV values, and one of the Portuguese employees has a very high HRV. It is worth stressing that workers in

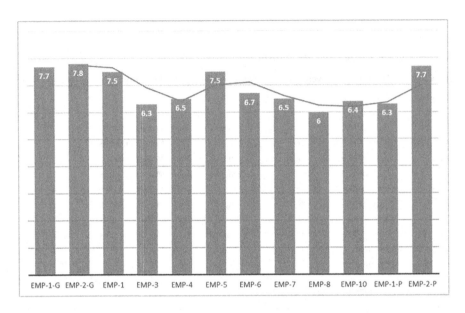

FIGURE 2.2   Average HRV values.

Italy displaced from eFM Headquarters to Masterandskills headquarters showed a decrease in their HRV values, which can be explained not only by the movement, but also by the different workplace conditions. As previously mentioned, eFM Headquarters is a role model in terms of good/excellent microclimate/environmental conditions and Masterandskills headquarters is located in an old building, with less favourable conditions.

If the HRV is lower, it can represent a higher level of stress. So we can deduce that the employees presenting a higher level of stress are emp-3, emp-4, emp-7, emp-8, emp-10 in Italy, and emp-1-P in Portugal. As these results represent the mean value of each employee's HRV, the variances between employees' HRV might not be related to higher or lower levels of stress, but to each specific person. Stress levels will be related with the variance of HRV to the HRV baseline of each worker.

Figure 2.3 shows that the employees in Greece have a higher HRV Threshold, a value that is compared to the mean values of employees in other countries. We can conclude that HRV depends on where the offices are located.

After the data collection phase, all information was processed and compared in order to understand the relationship between the HRV value and activities performed by each employee.

Thus, each employee had a minimum (lower limit) and a maximum HRV (upper limit). HRV depends on the activities in which the employee participates. Based on those limits, we calculated the average HRV and consider that as a "Threshold HRV". The input data were the average values of "Threshold HRV" for each employee and the average value of HRV for each activity, in the following terms:

Average HRV of each employee = (Lower limit of HRV + Upper limit of HRV)/2

$$(2.1)$$

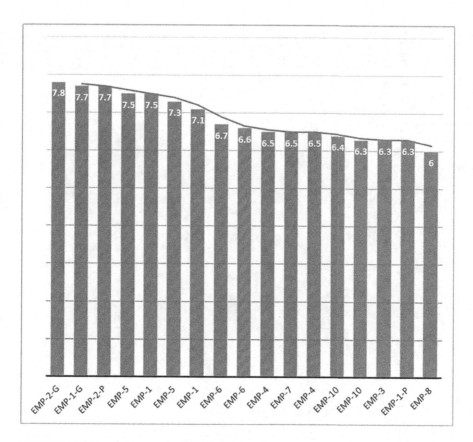

FIGURE 2.3   Decreasing range of HRV with location.

Table 2.2 shows the activities each employee performed during the monitoring phase. The highlighted values show the percentage correlation between the "Threshold HRV" and the average value of the individual activities. When this value is positive, we can identify the best well-being conditions for each employee and for each individual activity. Light grey indicates the activities/conditions that improve employees' well-being, as an increase of the HRV value in relation to the HRV threshold is observed.

In Table 2.3, the dark grey cells indicate the activities/conditions that are bad for employees, as a decrease in the HRV value in relation to the HRV threshold is noticed.

Considering the activity of "Presentation to other people", we can identify an association between the level of well-being and stress in the activity being performed. Table 2.4 shows clearly that the level of well-being was greater when the employees listened to other people's presentations than when they presented to others.

Table 2.5 shows that HRV can even depend on timing, from observing the subsequent Figures 2.4–2.6, where the HRV differs with the timeframe for each employee.

When analysing these data according to the location of the workplace, Figure 2.4 shows the HRV of the employees in Greece, which reveals a decrease in the HRV value when they return to work after the lunch break, and then it becomes steady after a short time. So it can be said that on returning to work employees display a brief period of higher-than-usual levels of stress.

TABLE 2.2

Activities That Increase Well-Being

Employee	Emp-1-G	Emp-2-G	Emp-1	Emp-3	Emp-4	Emp-5	Emp-6	Emp-7	Emp-8	Emp-10	Emp-1-P	Emp-2-P
Location	Greece	Greece	Italy	Italy	Italy	Italy	Italy	Italy	Italy	Italy	Portugal	Portugal
Telephone call-a	3%	1%	1%	3%		4%		7%				8%
Planning of the working day-b		1%										
Working alone (pc)-1		1%	1%			1.20%			0.50%	6%		15%
Working alone (to create something new)-2				9%		1.20%					1.50%	
Working under pressure-3	11%	3%	3%	3%					0.60%	5%		
Working under non-pressure-4					6%					5%		
Working alone (administrative duties)-5												15.60%
Working in small group-6			1%	10%						8%		8%
Conference / video call-c			1%							7%		
Printing documents-d	3%	1%										
Assigns tasks to employees-e				8%		4%		9%				
Controls work performed by other people-f					8.50%	4%		12%				
Debriefing with employees-g				3%	1.50%			7%				
Coordinating the team-h					1.50%							
Decision about something-i	2%		3%	10%								
Presentation to other people (speech)-j												
Presentation to other people (listening)-k			3%									
Talking about work-l	3%	1%	8%			1.20%		15%		5%		14.60%
Talking about other stuff-m										5%		4%
Lunch-n								3%		5%		11.30%
Coffee break-o												16.60%
Sporting activity-p												
Travel by car (to-from clients)-q												4.10%

Source: Own elaboration.

TABLE 2.3
Activities that Decrease Well-Being

Employee	Emp-1	Emp-2	Emp-1	Emp-3	Emp-4	Emp-5	Emp-6	Emp-7	Emp-8	Emp-10	Emp-1	Emp-2
Location	Greece	Greece	Italy	Italy	Italy	Italy	Italy	Italy	Italy	Italy	Portugal	Portugal
Telephone call-a												
Planning of the working day-b			−3%							−2%	−3.20%	
Working alone (pc)-1	−0.03			−5%	−2.50%						−1.50%	
Working alone (to create something new)-2									−0.40%	−5%		
Working under pressure-3					−12.50%	−1.30%						
Working under non-pressure-4	−0.02										−1.50%	
Working alone (administrative duties)-5		0										
Working in small group-6											−5%	−13%
Conference / video call-c				−5%	−20%							
Printing documents-d					−10%							
Assigns tasks to employees-e	−0.11											
Controls work performed by other people-f												
Debriefing with employees-g						−3%					−1.50%	
Coordinating the team-h					−2.40%	−3%					−3.20%	
Decision about something-i						−4%					−5%	
Presentation to other people (speech)-j		−0.04	−1%									−6%
Presentation to other people (listening)-k												
Talking about work-l											−1.50%	
Talking about other stuff-m			−3%		−1.60%							
Lunch-n												
Coffee break-o												
Sporting activity-p												
Travel by car (to-from clients)-q						−7%						

Source: Own elaboration.

TABLE 2.4
Variation of Well-Being Considering the Task of Presentation to Other People

Presentation to Other People (Speech)-J	Presentation to Other People (Listening)-K

-0.04
-1%

3%

-4%

5%

-6%   14.6%

Source: Own elaboration.

TABLE 2.5
HRV According to the Time of Day

Time	Emp-1-G	Emp-2-G	Emp-1-P	Emp-2-P	Emp-1	Emp-3	Emp-4	Emp-5	Emp-6	Emp-7	Emp-8	Emp-10
10.00 10.30	5.80	7.70	6.33	7.70	7.70	6.34	6.10	7.11	6.47	6.61	6.80	6.04
10.30 11.00	5.80	7.77	6.43	7.77	7.77	6.30	6.10	7.50	6.59	6.57	6.80	6.29
11.00 11.30	6.00	7.75	6.40	7.75	7.75	6.30	6.90	7.80	6.74	6.57	6.60	6.49
11.30 12.00	5.80	7.71	6.46	7.71	7.71	6.34	6.30	7.63	6.61	6.57	6.70	6.55
12.00 12.30	6.00	7.81	6.56	7.81	7.81	6.25	6.30	7.63	6.95	6.51	7.00	6.40
12.30 13.00	6.00	7.82	6.24	7.82	7.82	6.68	6.80	7.43	6.69	6.63	6.50	6.71
13.00 13.30	6.00	7.88	6.00	7.88	7.88	6.24	6.60	7.33	6.45	6.61	6.70	6.28
13.30 14.00	6.00	7.90	6.14	7.90	7.90	5.59	6.50	7.29	6.75	6.33	6.70	6.44
14.00 14.30	4.30	7.52	6.16	7.52	7.52	6.35	6.10	7.46	6.39	6.13	7.00	6.37
14.30 15.00	4.20	7.42	6.14	7.42	7.42	6.79	6.30	7.48	6.60	6.43	6.90	6.30

(*Continued*)

TABLE 2.5 (*Continued*)
HRV According to the Time of Day

Time	Emp-1-G	Emp-2-G	Emp-1-P	Emp-2-P	Emp-1	Emp-3	Emp-4	Emp-5	Emp-6	Emp-7	Emp-8	Emp-10
15.00 15.30	5.80	7.52	6.31	7.52	7.52	6.33	6.60	7.48	6.73	6.48	6.70	6.21
15.30 16.00	6.10	7.73	6.51	7.73	7.73	6.56	6.70	7.46	6.87	6.41	7.20	6.46
16.00 16.30	5.90	7.72	6.16	7.72	7.72	6.14	6.60	7.62	6.68	6.71	7.00	6.55
16.30 17.00	6.00	7.86	6.27	7.86	7.86	5.66	6.50	7.52	6.85	6.40	7.27	6.53
17.00 17.30	5.80	7.82	6.64	7.82	7.82	6.29	6.40	7.49	6.74	6.47	7.07	6.45
17.30 18.00	5.90	7.71	6.37	7.71	7.71	6.81	6.50	7.66	6.70	6.47	7.00	6.71

Source: Own elaboration.

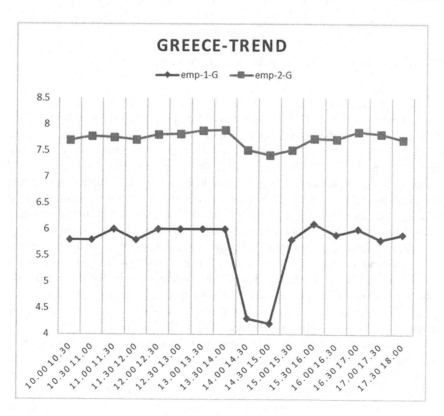

FIGURE 2.4 HRV of the employees in Greece, according to time.

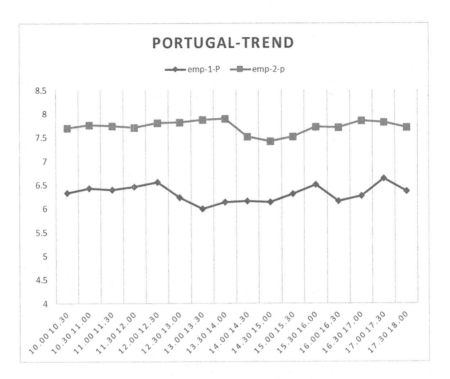

FIGURE 2.5   HRV of the employees in Portugal, according to time.

In Portugal, as shown in Figure 2.5, the two lines show a different trend. There is no correlation between the HRV value of the two employees during the working day from 10:30 to 17:00, while the trend is the same at the beginning and the end of the day. The graph shows how the trend of entry and exit reflects a common trend between the two employees: an increase of HRV values when they start work and a decrease of HRV values when they finish. So we can infer that at the beginning of the working day, as both workers' HRV values increase, the level of stress decreases. We can also assume that at the end of the day, as both workers' HRV values decrease, the level of stress increases, a possible explanation being that both workers rush to finish their jobs at the end of the day.

Figure 2.6 shows the trend lines of employees working in Rome. Each employee has their own trend line related to the time. There is no correlation between the different trends. During lunch time, the graph shows two different trends: Trend 1 (Emp-3–4–7–8) shows a low HRV, whereas Trend-2 (Emp-1–5–6–10) shows a high HRV.

Table 2.6 presents the activities which increase well-being, i.e. mainly causing an increase in the worker's HRV values, according to the location.

For example, concerning the Portuguese workers, making calls, working alone at the computer or doing administrative tasks, devoting their time to activities in small groups, listening to other colleagues' presentations, spending their daily periods talking with co-workers about work-related issues or other things, travelling by car to visit clients or having meetings with other stakeholders, or using breaks for resting are all activities that contribute to the employees' well-being, as seen in Table 2.6.

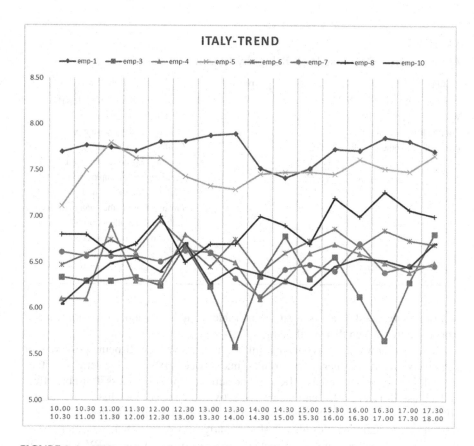

FIGURE 2.6   HRV of the employees in Italy, according to time.

TABLE 2.6
Activities That Increase Well-Being, by Country

Portugal	**Greece**	**Italy**
Telephone call-a	Telephone call-a	Telephone call-a
Working alone (pc)-1	Working under pressure-3	Working alone (pc)-1
Working alone (administrative duties)-5	Printing documents-d	Working under pressure-3
Working in small group-6	Talking about work-l	Working in small group-6
Presentation to other people (listening)-k		Assigns tasks to employees-e
Talking about work-l		Debriefing with employees-g
Talking about other stuff-m		Decision about something-i
Coffee break-o		Presentation to other people (listening)-k
Travel by car (to-from clients)-q		Talking about work-l

Source: Own elaboration.

TABLE 2.7
Activities That Decrease Well-Being, by Country

Portugal	Greece	Italy
Planning of the working day-B		Planning of the working day-B
Working in small group-6		Working alone (PC)-1
Coordinating the team-H		Working Under Pressure-3
Presentation to other people (speech)-J		Conference / video call-C
		Assigns tasks to employees-E
		Debriefing with employees-G
		Coordinating the team-H
		Presentation to other people (speech)-J
		Talking about other stuff-M
		Travel by car (to-from clients)-Q

Source: Own elaboration.

Conversely, the data analysis shown in Table 2.7 indicates a negative influence of workers' activities on their well-being.

Table 2.8 shows the activities that influence employees' well-being positively or negatively. In Greece, there is no straight link between HRV and workers' activities. Indeed, few activities increase HRV and no activities produce a less desirable effect (a decrease in HRV). This can mean that activities or conditions have no effect on HRV variance, so the environment and work organization are effective for workers' well-being. In Italy, employees are significantly affected by their work and often the same activities produce both increases and decreases in the HRV values. On the other hand, looking at the HRV results in the different Italian locations (eFM headquarters and Masterandskills headquarters), there are no significant variances. Portugal shows an intermediate situation, with a lot of activities improving the HRV index (increasing HRV, which means that stress is lower) and just some of them have a negative impact. Going deeper into single activities, we can see from the table that having a phone call is in general an activity with a positive effect on HRV values, irrespective of the location, as happens with working under pressure, talking about work and producing/printing documents. Making a plan of the working day, coordinating a team and making presentations to other people do not affect the HRV index positively, and in some periods, even have a negative impact on well-being. In fact, making presentations to an audience, interacting with colleagues or the public, induces stress and sometimes impacts negatively on the HRV index (decreasing the HRV values).

It was observed that some activities had a negative impact on HRV values, and thus on well-being, meaning that HRV values decrease when the worker is performing these activities. This negative influence was detected in planning the working day, having conference calls, coordinating the team, or making oral presentations to other people.

However, a great number of activities have a positive impact on workers' well-being, meaning that they influence HRV values positively, increasing them. Example

TABLE 2.8

Activities That Influence Well-Being (Positively in Light Grey and Negatively in Dark Grey)

Employee	Emp-1-g	Emp-2-g	Emp-1	Emp-3	Emp-4	Emp-5	Emp-6	Emp-7	Emp-8	Emp-10	Emp-1-p	Emp-2-p
Location	Greece	Greece	Italy	Italy	Italy	Italy	Italy	Italy	Italy	Italy	Portugal	Portugal
Telephone call-a	3%	1%	1%	3%		4%		7%				8%
Planning of the working day-b		1%	-3%							-2%	-3.20%	15%
Working alone (pc)-1		1%	1%	-5%	-2.50%	1.20%			0.50%	6%		
Working alone (to create something new)-2				9%		1.20%				-5%		
Working under pressure-3	11%	3%	3%	3%	-12.50%	-1.30%			0.60%	5%	1.50%	
Working under non-pressure-4					6%					5%		
Working alone (administrative duties)-5												15.60%
Working in small group-6			1%	10%	-20%					8%	-5%	8%
Conference / video call-c			1%	-5%						7%		-13%
Printing documents-d	3%	1%										
Assigns tasks to employees-e				8%	-10%	4%		9%				
Controls work performed by other people-f					8.50%	4%		12%				
Debriefing with employees-g				3%	1.50%	-3%		7%				
Coordinating the team-h						-3%					-3.20%	
Decision about something-i				10%	1.50%	-4%					-5%	
Presentation to other people (speech)-j	2%											-6%
Presentation to other people (listening)-k								15%		5%		14.60%
Talking about work-l	3%	1%	3%			1.20%				5%		4%
Talking about other stuff-m			8%							5%		11.30%
Lunch-n			-3%									
Coffee break-o								3%				16.60%
Sporting activity-p												4.10%
Travel by car (to-from clients)-q						-7%						

Source: Own elaboration.

of this type of activities are making professional telephone calls, working under pressure, working in small groups, producing/printing documents, assigning tasks to employees, listening to presentations and talking about work.

2.5 CONCLUSIONS, LIMITATIONS AND FUTURE RESEARCH

This chapter analyses, in an innovative way, how (and to what extent) the environmental and microclimatic conditions, as well as the organization of employees' daily routines and tasks, impact on workers' well-being, using a measure based on HRV.

Regarding the relation between HRV and timing, it is outlined that the input and output values tend to be low, and the average trend detected corresponds to (1) increased HRV, when workers start work, and (2) decreased HRV, when they stop working. The HRV value tends to be low due to stress related to reaching the office, while the HRV value tends to be high immediately after that stress decreases. At the same time, the HRV value tends to be low due to stress related to leaving the office, in completing the last activity of the working day. A possible solution to mitigate such effects could be the implementation of smart working, including innovative ways of work organization, such as tasks prioritization, lean management procedures, design thinking approaches, etc.

As stated before, if the HRV is lower, it can represent a higher level of stress and the microclimatic/environmental conditions related to the location can influence the worker's well-being. The employees presenting a higher level of stress are emp-3, emp-4, emp-7, emp-8 and emp-10 in Italy and emp-1 in Portugal. Moreover, it is worth to stress that moving workers in Italy from eFM headquarters to Masterandskills headquarters showed a decrease in their HRV values. This may be due not only to the displacement effect, but also to the different conditions in these places. As mentioned earlier, eFM headquarters has excellent microclimatic/environmental/layout conditions, contrasting with Masterandskills headquarters which is an old building, with poor ergonomic conditions, poor lighting and the lack of an innovative layout.

Regarding work organization in terms of workers' activities and HRV, a set of activities contribute to the Portuguese workers' well-being, namely making calls, working alone at the computer or doing administrative tasks, devoting their time to activities in small groups, listening to other colleagues' presentations, spending periods talking with co-workers about work-related issues or other things, travelling by car to visit clients or having meetings with other stakeholders or using break periods for resting.

For Greek employees, we found no direct linkage between HRV and their activities. Moreover, we detected few activities that increase HRV. Thus, a concluding remark that deserves to be outlined is that the activities or conditions have no significant effect on HRV variance, so these workers' environment and work organization processes contribute to their well-being.

In turn, for Italian employees, their work activities and routines had a significant influence on HRV values. Furthermore, this same set of activities frequently produces both increases and decreases in HRV. These work activities and routines do not differ significantly when contrasting the two Italian locations (eFM headquarters and Masterandskills headquarters).

The empirical findings now obtained reinforce also the need for incorporating additional measurement parameters, besides data monitoring based on HRV, unveiling a set of substantial effects on workers' well-being. The current analysis could be improved using a more significant number of clusters. Therefore, HRV may not be the only indicator of well-being changing during the day; thus, collecting an additional set of indicators is required.

This chapter has limitations regarding the need for other parameters of measurement that can improve the quality of results. For instance, taking into account the personal conditions related to employees' private life and personal behaviour may influence HRV. HRV variability is one key indicator of stress, but in order to describe a more realistic stressful condition, it should be integrated with other parameters, possibly chemical factors such as salivary cortisol, adrenaline, noradrenaline, heart rate, sudation and others. One more limitation is related to the fact that every employee reacts differently to the same stress level, something that must be taken into consideration. The sample is limited due to the small number of employees, and so it could be improved by adding more workers and extending the duration of the study. In addition, it covers a very specific workplace typology or activity sector, i.e. services, mainly characterized by consultancy and management activities. Further research efforts could explore other private and public workplaces, having different organizational layouts and environmental conditions.

In the scope of the current research endeavour, it was not possible to measure temperature, air quality, indoor airflow and lighting through sensors in the workplace due to the lack of financial resources. Nevertheless, these qualitative data were recorded by both trainees and employees.

In terms of future research, possible improvements can be achieved by expanding the time period, including different cohorts. Also, by observing the above results, it is concluded that each employee has their own mindset/HRV. In order to have a generalized pattern in HRV, it could be considered an additional number of employees in order to improve the empirical findings and generalize the results.

This research endeavour can be further extended by considering different locations because each country has its own working environment and the mindset of the employer (organization) is different. Therefore, considering different countries will probably provide different insights, contrasts and implications. Also, bearing privacy issues in mind, if the employee's health is taken into consideration, the HRV patterns could be differentiated. For example, an employee suffering from anxiety/depression/high or low blood pressure will show a different HRV from a healthy employee.

Finally, asking every employee about their satisfaction with their work/boss will play a key role in determining the results. A satisfied employee will have a stable HRV even in stressful situations compared to a dissatisfied employee. Adding different departments to this project could also contribute for obtaining new evidences, since not every employee will participate in the same activities/conditions all the time.

FUNDING

The Portuguese Foundation for Science and Technology (Grants and NECE-UIDB/04630/2020) provided support for this study.

ACKNOWLEDGEMENTS

The authors acknowledge the highly valuable comments and suggestions provided by the editor and reviewers, which contributed to the improvement in the clarity, focus, contribution, and scientific soundness of the current study.

CONFLICTS OF INTEREST

The authors declare there are no conflicts of interest.

REFERENCES

Acharya, U.R., Joseph, K.P., Kannathal, N., Lim, C.M., and Suri, J.S. 2006. Heart rate variability: A review. *Medical and Biological Engineering and Computing* 44 (12): 1031–51. Doi: 10.1007/s11517-006-0119-0.

Aked, J., Marks, N., Cordon, C., and Thompson, S. 2009. *Five Ways to Wellbeig: A Report Presented to the Foresight Project on Communicating the Evidence Base for Improving People's Well-Being.* Centre for Well-Being, NEF (the New Economics Foundation). Doi:10.7748/ns2013.04.27.34.29.s38.

Al horr, Y.A., Arif, M., Katafygiotou, M., Mazroei, A., Kaushik, A., and Elsarrag, E. 2016. Impact of indoor environmental quality on occupant well-being and comfort: A review of the literature. *International Journal of Sustainable Built Environment* 5 (1): 1–11. Doi:10.1016/j.ijsbe.2016.03.006.

Aryal, A., Parish, M., and Rohlman, D. 2019. Generalizability of total worker health® online training for young workers. *International Journal of Environmental Research and Public Health* 16 (4): 577. Doi: 10.3390/ijerph16040577.

Aziri, B. and Irfan, M. 2011. Job satisfaction: A literature review. *Management Research and Practice* 3 (4): 77–86.

Bailon, R., Laguna, P., Mainardi, L., and Sornmo, L. 2007. Analysis of heart rate variability using time-varying frequency bands based on respiratory frequency. *Conference Proceedings: ... Annual International Conference of the IEEE Engineering in Medicine and Biology Society. IEEE Engineering in Medicine and Biology Society. Conference,* 6675–78.

Bakker, A.B. 2015. Towards a multilevel approach of employee well-being. *European Journal of Work and Organizational Psychology* 24 (6): 839–43. Doi: 10.1080/1359432X.2015.1071423.

Bakotić, D. and Tomislav, B. 2013. Relationship between working conditions and job satisfaction : The case of croatian shipbuilding company. *International Journal of Business and Social Science* 4 (2): 206–13.

Berman, M.G, Jonides, J., and Kaplan, S. 2008. The cognitive benefits of Interacting with nature. *Psychological Science* 19 (12). SAGE Publications Inc: 1207–12. Doi: 10.1111/j.1467-9280.2008.02225.x.

Boehm, J.K. and Lyubomirsky, S. 2008. Does happiness promote career success? *Journal of Career Assessment* 16 (1). SAGE Publications Inc: 101–16. Doi: 10.1177/10690 72707308140.

Burge, S., Hedge, A., Wilson, S., Bass, J.H., and Robertson, A. 1987. Sick building syndrome: A study of 4373 office workers. *The Annals of Occupational Hygiene* 31 (4): 493–504. Doi: 10.1093/annhyg/31.4A.493.

Carayon, P. and Smith, M.J. 2000. Work organization and ergonomics. *Applied Ergonomics.* Doi: 10.1016/S0003-6870(00)00040-5.

Carlopio, J.R. 1996. Construct validity of physical work environment satisfaction questionnaire. *Journal of Occupational Health Psychology* 24: 579–601.

Cohen, L.M. 2007. Bridging two streams of office design research: A comparison of design/ behaviour and management journal articles from 1980–2001. *Journal of Architectural and Planning Research* 24 (4). Locke Science Publishing Company, Inc.: 289–307. http://www.jstor.org/stable/43030809.

Cooper, S.C.L. and Leiter, M.P. 2017. *The Routledge Companion to Well-Being at Work*. London: Routledg.

Day, A., and Randell, K.D. 2014. Building a foundation for psychologically healthy workplaces and well-being. In *Workplace Well-Being : How to Build Psychologically Healthy Workplaces*. Chichester: John Wiley & Sons, Ltd

Dewe, P., and Cooper, C. 2012. *Well-Being and Work: Towards a Balanced Agenda*. Basingstoke: Palgrave Macmillan.

Dodge, R., Daly, A., Huyton, J., and Sanders, L. 2012. The challenge of defining wellbeing. *International Journal of Wellbeing* 2 (3): 222–35. Doi: 10.5502/ijw.v2i3.4.

Edwards, J.A., Van Laar, D., Easton, S., and Kinman, G. 2009. The work-related quality of life scale for higher education employees. *Quality in Higher Education* 15 (3): 207–19. Doi: 10.1080/13538320903343057.

Emrah, O., Osman, U., and Oguzhan, O. 2014. Who are happier at work and in life ? public sector versus private sector : A research on Turkish employees. *International Journal of Recent Advances in Organizational Behaviour and Decision Sciences* 1 (2): 148–60.

Ford, M.T., Cerasoli, C.P., Higgins, J.A., and Decesare, A.L. 2011. Relationships between psychological, physical, and behavioural health and work performance: A review and meta-analysis. *Work and Stress* 25 (3): 185–204. Doi: 10.1080/02678373.2011. 609035.

Gou, Z. 2019. Human factors in green building: Building types and users' needs. *Buildings* 9 (1): 17. Doi: 10.3390/buildings9010017.

Hackman, J.R. and Oldham, G.R. 1976. Motivation through the design of work: Test of a theory. *Organizational Behavior and Human Performance* 16 (2): 250–79. Doi: 10.1016/0030–5073(76)90016-7.

Hanaysha, J. and Tahir, P.R. 2016. Examining the effects of employee empowerment, teamwork, and employee training on job satisfaction. *Procedia - Social and Behavioral Sciences* 219: 272–82. Doi:10.1016/j.sbspro.2016.05.016.

Hartig, T., Evans, G.W., Jamner, L.D. Davis, D.S., and Gärling, T. 2003. Tracking restoration in natural and urban field settings. *Journal of Environmental Psychology*. Hartig, Terry: Institute for Housing and Urban Research, Uppsala University, Box 785, Gavle, Sweden, S-807 29, terry.hartig@ibf.uu.se: Elsevier Science. Doi: 10.1016/ S0272-4944(02)00109-3.

Holick, M.F. 2004. Sunlight and vitamin D for bone health and prevention of autoimmune diseases, cancers, and cardiovascular disease. *The American Journal of Clinical Nutrition* 80 (6). Doi:10.1093/ajcn/80.6.1678S.

Hone, L.C., Jarden, A., Duncan, S., and Schofield, G.M. 2015. Flourishing in New Zealand workers: Associations with lifestyle behaviors, physical health, psychosocial, and work-related indicators. *Journal of Occupational and Environmental Medicine* 57 (9): 973–83. Doi: 10.1097/JOM.0000000000000508.

IAEA, and ILO. 2018. Occupational radiation protection. Vienna. https://www-pub.iaea.org/ MTCD/Publications/PDF/PUB1785_web.pdf.

Im, S., Chung, Y., and Yang, J. 2018. The mediating roles of happiness and cohesion in the relationship between employee volunteerism and job performance. *International Journal of Environmental Research and Public Health* 15 (12): 2903. Doi: 10.3390/ ijerph15122903.

Ji, Z., Pons, D. and Pearse, J. 2018. Measuring industrial health using a diminished quality of life instrument. *Safety* 4 (4): 55. Doi: 10.3390/safety4040055.

Kahneman, D., Krueger, A.B., Schkade, D.A. Schwarz, N., and Stone, A.A. 2004. A survey method for characterizing daily life experience: The day reconstruction method. *Science* 306 (5702): 1776–80. Doi: 10.1126/science.1103572.

Keeman, A., Näswall, K. Malinen, S., and Kuntz, J. 2017. Employee wellbeing: Evaluating a wellbeing intervention in two settings. *Frontiers in Psychology*. 8: 1–14.

Kelloway, E.K. and Day, A. 2005. Building healthy workplaces: Where we need to be. *Canadian Journal of Behavioural Science/Revue Canadienne Des Sciences Du Comportement* 37 (4): 309–12. Doi: 10.1037/h0087265.

Keyes, C.L.M. and Grzywacz, J.G. 2005. Health as a complete state: The added value in work performance and healthcare costs. *Journal of Occupational and Environmental Medicine* 47 (5): 523–32. Doi: 10.1097/01.jom.0000161737.21198.3a.

Knight, C. and Alexander Haslam, S. 2010. The relative merits of lean, enriched, and empowered offices: An experimental examination of the impact of workspace management strategies on well-being and productivity. *Journal of Experimental Psychology. Applied* 16 (2): 158–72. Doi: 10.1037/a0019292.

Kottwitz, M.U., Schade, V., Burger, C., Radlinger, L., and Elfering, A. 2018. Time pressure, time autonomy, and sickness absenteeism in hospital employees: A longitudinal study on organizational absenteeism records. *Safety and Health at Work* 9 (1): 109–14. Doi: 10.1016/j.shaw.2017.06.013.

Kwon, O., Lim, S., and Lee, D.H. 2018. Acquiring startups in the energy sector: A study of firm value and environmental policy. *Business Strategy and the Environment* 27 (8): 1376–84. Doi: 10.1002/bse.2187.

Leather, P., Pyrgas, M., Beale, D., and Lawrence, C. 1998. Windows in the workplace: Sunlight, view, and occupational stress. *Environment and Behavior* 30 (6): 739–62. Doi: 10.1177/001391659803000601.

Lee, S.Y. and Brand, J.L. 2005. Effects of control over office workspace on perceptions of the work environment and work outcomes. *Journal of Environmental Psychology* 25 (3): 323–33. Doi:10.1016/j.jenvp.2005.08.001.

Leech, J.A., Nelson, W.C. Burnett, R.T. Aaron, S., and Raizenne, M.E. 2002. It's about time: A comparison of Canadian and American time-activity patterns. *Journal of Exposure Analysis and Environmental Epidemiology* 12 (6): 427–32. Doi: 10.1038/sj.jea.7500244.

Leitão, J., Pereira, D., and Gonçalves, Â. 2019. Quality of work life and organizational performance : Workers ' feelings of contributing, or not, to the organization's productivity. *International Journal of Environmental Research and Public Health* 16 (3803): 1–18.

Lyubomirsky, S., King, L., and Diener, E. 2005. The benefits of frequent positive affect: Does happiness lead to success? *Psychological Bulletin* 131 (6): 803–55. Doi:10.1037/0033-2909.131.6.803.

Maghsoodi, A.I., Azizi-Ari, I., Barzegar-Kasani, Z., Azad, M., Zavadskas, E.K., and Antuchevičienė, J. 2018. Evaluation of the influencing factors on job satisfaction based on combination of PLS-SEM and F-MULTIMOORA approach. *Symmetry* 11 (1): 24. Doi: 10.3390/sym11010024.

Marmaras, N., and D. Nathanael. 2006. *Workplace Design. Handbook of Human Factors and Ergonomics*. Wiley Online Books. Doi: 10.1002/0470048204.ch22.

Nanni, S., Benetti, E., and Mazzini, G. 2017. Indoor monitoring in public buildings: Workplace wellbeing and energy consumptions. an example of iot for smart cities application. *Advances in Science, Technology and Engineering Systems* 2 (3): 884–90. Doi: 10.25046/aj0203110.

Nieuwenhuis, M., Knight, C., Postmes, T., and Haslam, S.A. 2014. The relative benefits of green versus lean office space: Three field experiments. *Journal of Experimental Psychology: Applied*. Doi: 10.1037/xap0000024.

Raziq, A. and Maulabakhsh, R. 2015. Impact of working environment on job satisfaction. *Procedia Economics and Finance* 23: 717–25. Doi: 10.1016/S2212–5671(15)00524-9.

Ryan, R.M. and Deci, E.L. 2001. On happiness and human potentials: A review of research on hedonic and eudaimonic well-being. *Annual Review of Psychology* 52: 141–166.

Salonen, H., Lahtinen, M. Lappalainen, Sanna. Nevala, N., Knibbs, L.D., Morawska, L., and Reijula, K. 2013. Physical characteristics of the indoor environment that affect health and wellbeing in healthcare facilities: A review. *Intelligent Buildings International* 5 (1): 3–25. Doi: 10.1080/17508975.2013.764838.

Serghides, D.K., Chatzinikola, C.K., and Katafygiotou, M.C. 2015. Comparative studies of the occupants' behaviour in a university building during winter and summer time. *International Journal of Sustainable Energy* 34 (8): 528–51. Doi: 10.1080/14786451.2014.905578.

Sirgy, M., Efraty, D., Siegel, P., and Lee, D.A. 2001. New measure of quality of work life (QOWL) based on need satisfaction and spillover theories'. *Social Indicators Research* 55 (3): 241–302.

Stein, P.K., Bosner, M.S., Kleiger, R.E., and Conger, B.M. 1994. Heart rate variability: A measure of cardiac autonomic tone. *American Heart Journal* 127 (5): 1376–81. Doi: 10.1016/0002-8703(94)90059-0.

Velarde, M.D., Fry, G., and Tveit, M. 2007. Health effects of viewing landscapes - landscape types in environmental psychology. *Urban Forestry and Urban Greening* 6 (4): 199–212. Doi: 10.1016/j.ufug.2007.07.001.

Vischer, J.C. 2007. The effects of the physical environment on job performance: Towards a theoretical model of workspace stress. *Stress and Health* 23 (3): 175–84. Doi: 10.1002/smi.1134.

WHO. 2012. Health indicators of sustainable jobs. http://www.who.int/hia/green_economy/indicators_jobs.pdf.

WHO. 2016. Ionizing radiation, health effects and protective measures. https://www.who.int/news-room/fact-sheets/detail/ionizing-radiation-health-effects-and-protective-measures.

3 Quality of Life in Nursing Homes

A European Approach for Improvement

Willeke van Staalduinen
AFEdemy

Sylvie Schoch
IP-International GmbH

Karin Stiehr
ISIS GmbH

Javier Ganzarain
AFEdemy

Laura Annella
CADIAI Cooperativa Sociale

Rasa Naujaniene
Vytautas Magnus University

Eglé Gerulaitiene
Vytautas Magnus University

CONTENTS

3.1   INTRODUCTION

The overwhelming majority of older adults in Europe live in private households. Nevertheless, some older adults move into institutional households, such as residential care or nursing homes. This may occur out of choice (for example, not wishing to live alone) or because it is no longer possible for them to carry on living at home (for example, due to complex long-term care needs, such as the complications of dementia). In 2011, 3.8% of older women (aged 65 years or more) and 1.9% of older men in the EU-28 were living in an institutional household (Eurostat, 2019).

Present-day nursing homes[1] are places where people live to be reactivated after a major accident or hospital treatment and/or are offered a home with 24/7 support and care. The concept of nursing homes has mainly been developed in the 1960s and the 1970s of the past century to reduce the pressure on the occupancy of hospital beds. Hospitals in that time were faced with a growing number of patients (mainly older people) that could not return home and continued to occupy hospital beds for many months. Beds that were not equipped to deliver care to people with chronic geriatric diseases or impairments. To find alternative and qualitatively better solutions, nursing homes were build and at the start they mainly provided services according to the medical services model. This history still influences the current view of most elderly care organizations. However, this view slowly alternates towards the vision that nursing homes must be places where people live and feel at home.

Quality of life in our definition is the well-being of individuals, encompassing negative and positive features of life. Quality of life is subjective, culturally and individually defined, independent from occurring functional disabilities or living circumstances. In the scientific discourse, a widely shared understanding of quality of life in old age is that it is realized in the interaction between a person and their environment, depends substantially on the resources and potentials of the person concerned and can be promoted by positive framework conditions. The influence of age on the quality of life, however, is issue of debates. Compared to personal independence and social relationships, Kratzer (2011) considers age as an ineffective factor. Wiese (2015), on the other hand, described self-sufficiency in contrast to depression and multi-morbidity as important factors influencing quality of life, whereas marital status, socioeconomic status, education, nutrition and ethical affiliation tend to play a subordinate role. A consistent orientation towards the paradigm of active ageing takes place in studies and projects that focus on aspects of learning and commitment to the community – the right to engage in meaningful activities – also in old age and in the event of functional limitations.

---

[1] In this chapter, we use the overall term nursing homes to include all kinds of facilities for longer-term residential care, such as care homes, residential care homes and nursing hospitals.

In what ways quality of life of older people living in nursing or care homes is taken care of, and how this can be improved in four European countries (Germany, Italy, Lithuania and the Netherlands) by developing a compendium and a serious game, is the focus of this chapter. At first, we will look at nursing home facilities in these countries and the quality of care and then turn to the paradigm shift from quality of care towards quality of life.

3.2   NURSING HOME CARE IN EUROPE

The World Health Organization (WHO) and Milbank Memorial Fund outline important key points for dealing with older people in need of care. They define long-term care as follows:

> Long-term care is the system of activities undertaken by informal caregivers (family, friends, and/or neighbours) and/or professionals (health, social, and others) to ensure that a person who is not fully capable of self-care can maintain the highest possible quality of life, according to their individual preferences, with the greatest possible degree of independence, autonomy, participation, personal fulfilment, and human dignity.

(WHO, 2000)

While most healthcare costs in the European Union are covered by social protection systems, long-term social care is usually not covered to the same extent as healthcare. This means that the responsibility for financing institutional care often resides with the older person needing such care (or with their family) (Eurostat, 2019). Long-term care in Germany is provided within the institutional framework of long-term care insurance that is mandatory for every citizen. In Italy, long-term care is characterized by a wide variation among regions and areas in both funding levels and the structure of the services provided. Long-term care in Lithuania is a new and developing area of social policy. In the Netherlands, residential care is funded by a collective insurance scheme with income-related contributions, to which all Dutch citizens have access. (ESPN, 2018; GAMLEC, 2020).

In Germany, a development towards small-structured entities is being observed. Nursing homes until the 1960s were holding institutions with big wards, a high occupancy rate and no or little privacy for "inmates". In the 1960s and 1970s, nursing homes resembled hospitals with still up to three "patients" per room. In the 1980s and 1990s, big facilities were segregated into smaller communities with the aim of activating care in a more personal environment, and dependent older people are since then called "residents". They are often provided with accompanying services, such as hairdressers within or close to the facility. As of the middle of the 1990s, shared flats and houses are on the rise with care provision like in a family. However, the overwhelming majority of residents nowadays live in a facility of the third generation created in the 1980s and 1990s. The allowed maximum occupancy rate are two persons in one room (KDA).

In Italy, nursing homes are generally organized into organizational units for the provision of residential services named "nuclei". Each "nucleo" is an autonomous area in terms of equipment and services, usually consisting of 20 beds (in single or multiple bedrooms), one living room and one dining room. In addition, in the nursing homes, there are areas open to all residents, dedicated to leisure, beauty or worship activities and rehabilitation.

Depending on their level of dependency and care needs, disabled people in Lithuania may receive permanent home care (assistance provided for recipients that continue living in their own home) or permanent nursing care in an institutional setting. Long-term care in the health sector is mostly provided as inpatient care in specialized nursing hospitals or in specific departments in general hospitals.

In the Netherlands, similar developments occur as in Germany. Since the 1970s of the past century, nursing homes were mainly organized in wards of about 15 residents with single or multiple bedrooms and one or two common living rooms. Additionally, they provide facilities such as restaurants, cafes, hairdresser, therapy rooms, shops and rooms for prayer. Over the past several decades, nursing homes in the Netherlands also offer care in smaller group accommodations of 6–8 residents to provide a home like a family to residents. Another development is the increase of nursing home care supply at home.

The paid caring and nursing staff of nursing homes consist of care assistants, caregivers and nurses (HHM, 2015). They take care of the daily care delivery, varying from showering and getting dressed till the supply of medicines. Medical or paramedic staff of nursing homes consists of doctors, physical activists and psychologists. Dieticians, social work and housekeeping complete the usual paid staff in nursing homes in Europe. Additionally, paid staff is being supported by volunteers and family, who provide, for example, in hostess functions, such as coffee meetings or support residents in participation activities.

The majority of nursing home residents in Europe are aged over 80. Dementia (mainly Alzheimer's disease and vascular dementia), non-congenital brain injury or damage and severe respiratory conditions are the main causes for nursing home admission. Their care needs vary from receiving support in case of severe memory loss, severe incontinence, mobility problems, behavioural problems and artificial respiration. These health issues cannot be addressed anymore by the persons themselves or their families and informal care takers. They need care and supervision around the clock (24 hours, 7 days per week) which is mainly offered by nursing homes. In most cases, nursing home residents are completely dependent from others in their daily activities such as getting dressed, eating, bathing or walking.

3.3   QUALITY OF CARE IN NURSING HOMES AND EUROPEAN ANSWERS

For several years, nursing home facilities have been faced with the unsolved problem of finding a sufficient number of qualified nursing staff and caregivers. International studies show that specialists in geriatric care are essential in order to offer residents quality of care and to ensure an adequate level of quality of life in the facility. Working conditions in the nursing sector are generally more stressful than in other areas of work in terms of physical and mental requirements (Theobald, 2013; Neldner, 2017). Shift work, the use of agency work, high work intensity and suboptimal work organization result in high fluctuation and in many cases even the exit from the profession (Klein & Gaugisch, 2005). Despite extra financial efforts, such as 2 billion euro extra for staff in the Netherlands in 2021, it is hard to find qualified and motivated personnel and to keep them.

In 2015, also the WHO reported that the quality of long-term care often leaves much room for improvement, even in high-income countries. Quality is undermined by two major factors: the type of care that is provided and the lack of effective regulations and standards, as well as the low priority given to long-term care. Although there are outstanding exceptions, significant threats to the quality of care origin from outdated ideas and ways of working, which often focus on keeping older people alive rather than on supporting dignified living and maintaining their intrinsic capacity (WHO, 2015).

In 2018, the European Social Policy Network (ESPN, 2018) concluded that long-term care in Europe faces several challenges. There is the challenge of insufficient availability of residential care due to structural factors such as deinstitutionalization and funding. Quality of care is another challenge, also in relation to the working conditions, that remain low and are depicted negatively, such as low income, lack of training, high workload and high level of strain. This leads to a severe shortage of qualified professionals.

In Germany, as of November 2019, revised measures to assess and document the quality of care in residential settings became effective. Public reporting is based on (1) general information about the facility, (2) quality data collected by care homes on predefined criteria and (3) random quality checks by the Medical Service of Health Insurance Funds. Nursing care facilities have to collect the data every 6 months and forward them to an evaluation centre. Data is also provided on topics that are relevant not only for the quality of care but also cover aspects of the quality of life of care home residents.

Improving the quality of care in nursing homes is a compelling goal of the Italian health care system. The implementation of it is complex and articulated and cannot be separated from moving in different directions in line with socio-demographic changes, regional differences, and the progress of knowledge and scientific debate. Italian regions and autonomous provinces use programmes such as accreditation, a form of external assessment of quality and safety, to assess, promote and maintain high quality and safe services. Minimum standards for the authorization and accreditation of nursing homes include requirements such as accessibility, the employment of a coordinator and qualified professionals (specified by regional authorities in terms of staffing levels), individual care planning and publication of a Service Charter. Regional variations are therefore inevitable but they result in relatively similar regulations. Local health authorities promote systematic audit activities to enhance self-assessment of professionals and improve clinical practice. Moreover, since quality services mainly depend on well-trained professionals, regional accreditation systems require to provide continuous training of care workers and all paid staff (GAMLEC, 2020).

In Lithuania, social care in residential care homes for older people is regulated according to standards developed by the Ministry of Social Security and Labour. During the last couple of years, the European Quality in Social Services (EQUASS) system was introduced in some care homes, applying Voluntary European Quality Framework for Social Services (European Commission, 2016).

To deliver a minimum of quality of care to nursing home residents, the Dutch government provided a Quality Framework in 2017, supervised by the Health Inspectorate. Quality of care provides person-centred care and support to the resident, taking care of the resident's personal history, future and goals. This takes place within the relationship between resident, family, care professional and care organization. The quality of this relationship is one aspect of quality of care. Second aspect is the quality of housing and well-being and how professionals and the care organization take care of that. Thirdly, quality of care guarantees the safety of residents according to professional standards and guidelines and to avoid damage to residents. Final aspect is the way care professional and organization provide optimal care to residents and continue to learn (Kwaliteitskader, 2017).

3.4 THE PARADIGM SHIFT FROM QUALITY OF CARE TO QUALITY OF LIFE

In the debate of long-term care, a paradigm shift from quality of care to quality of life has been made over time. Quality of care is often valued as too narrow, and mainly consisting of care delivery and hygiene norms, where quality of life is focusing on physical, psychological and social well-being of residents. However, still it is essential that provided (medical) care has a high level of quality; otherwise, it inevitably affects the quality of life of residents.

Defining and promoting the quality of life of nursing home residents has to cope with two challenges:

1. Unlike quality of care, which can be assessed by measurable indicators, quality of life is characterized by soft factors which for each individual vary in importance.
2. If quality of life in nursing homes is to be promoted, the focus must be on the introduction of structural conditions, especially the offer of opportunities for individuals to realize their personal quality of life.

A European partnership of organizations from four European countries: Germany, Italy, Lithuania and the Netherlands, was created in 2019 to find innovative ways to open up the learning and discussions on quality of life in nursing homes. This consortium of the European Erasmus+ project Gaming for Mutual Learning in Elder Care (GAMLEC 2019–2021) consists of a multidisciplinary group of workers from nursing homes, social scientists, computer scientists and trainers with the aim to jointly build a serious board game. The reason to choose for the methodology of serious gaming is that several researches identify that serious games provide users with valuable information in a fun and entertaining way (Connolly, 2012; Hamari, 2014). Learning by using serious games in the healthcare sector offers workers to develop skills in a safe and learner-oriented environment and with less costs (Odessa, 2013; Ricciardi, 2014). This potentially engages and inspires users more than traditional methods of learning. The GAMLEC consortium expects that learning on the topic of quality of life best can be addressed if the players are invited to learn in an inspiring, easily accessible and playful way.

Regardless of the approach chosen, **autonomy, participation and human dignity** are the cornerstones of dignified ageing according to the WHO (2000). These WHO dimensions create the framework for determining factors for quality of life of older people who are not self-sufficient. The GAMLEC partners performed desk researches and interviews on autonomy, participation and human dignity in their countries. These analyses have been used to create the European Compendium, which in turn will be translated into the board game for nursing home staff, volunteers, residents and family. In the following pages, the cornerstones of quality of life will be further explored and the game will be introduced.

3.4.1 INTRODUCTION OF THE COMPENDIUM

The GAMLEC Compendium is based on research on widespread deficiencies in institutionalized long-term care and proposes 68 criteria according to which the quality of life of nursing home residents can be ensured. In the following section, problematic areas and exemplary solutions are described to enhance the autonomy, participation and human dignity of nursing home residents. As a rule, these solutions are not requiring less financial investments than rather a change in attitude of persons in relation to dependent older people.

Cultural differences between countries in Europe are mainly found in institutionalized long-term care systems, for example, in terms of available places or their funding mechanisms. But these differences are dwarfed by a shared comprehension of basic differences between "us" and "them", defining dependent older people as a kind of somehow different species. For them self-evident preconditions for quality of life – such as making one's own decisions, being respected, engage in meaningful activities – are put in doubt or even declared inapplicable.

The personal role that is taken in institutionalized care settings is decisive for the apprehension of dependent older people's quality of life.

The most important stakeholders in this context are:

- **Nursing home managers**: They are in biggest distance to people in need of care as they have to run an organization.
- **Professional caregivers**: Influenced by their vocational training and working conditions, their predominant focus is on the physical well-being of dependent old people.
- **Family members and other persons of trust, such as volunteers**: They have the closest relation to care home residents and can function as their advocates. However, their judgement is often dulled by feelings of guilt or helplessness when the situation of the old person deteriorates.
- **Dependent older people**: Taken into account the above-mentioned constraints, they must determine the preconditions of their quality of life and are perfectly able to express them even in a late phase of dementia.

3.4.2 AUTONOMY

When queried on their attitudes to balancing risk and autonomy for residents with dementia, English care home managers outlined three conflicting areas, in which individual's needs must be subjugated to primary goals: ensuring safety by the provision of limited access to outdoor space, protecting the dignity of residents with dementia and achieving a balance with the needs of other residents (Evans, 2018).

From the perspective of nurses, as reported in a Swedish study, the awareness of older people's frailty in nursing homes and the importance of maintained health and well-being were described as the main source for promoting autonomy and participation (Hedman, 2017). Older people themselves have a differentiated view on this topic. In a German survey on aspects of quality of life in old age, health in the sense of mental and physical well-being was indeed considered crucial in general. For a strong minority of nearly 40%, however, health was not necessarily most important. Harmonic relationships with others, above all, with a permanent partner, spirituality, contentment and the feeling of being at peace with one another, are powerful factors for the quality of life that can put the meaning of health into perspective. In particular, a higher level of education seems to lead to health not being considered the most important prerequisite for quality of life (Stiehr, 2016).

Autonomy includes personal independence and freedom of will in one's actions and refers to the experience that one can choose activities, make decisions and behave in accordance with one's goals (Custers et al., 2012). With functional restrictions or limited health, personal independence and freedom of will need to be assisted in order to be realized.

This collides in many ways with institutional settings with impacts on the personal routines, as reported from all partner countries. Life in nursing homes is organized according to pre-set time schemes, personal preferences as regards times for sleeping, eating and personal hygiene are seldom respected: "Most people get up before breakfast, the food is the same, the canteen doesn't offer a choice from several dishes. You can smoke only in the designated area" (Vice-manager in care facility, Klaipeda region, Lithuania).

Shortages in staff prevent person-centred care relations, and overburdened caregivers are at risk to exert dominance on vulnerable persons who cannot defend themselves. Residents suffer from the lack of freedom in maintaining some daily habits such as: to take a walk in the neighbourhood or to make personal purchases by themselves. But:

> Many nurses would like to take more care of the individual concerns of the residents. It is the task of the facility manager that their time budget allows for that. A good duty-roster is the first step to make it possible (Employee of Heimverzeichnis GmbH, Frankfurt am Main, Germany).

Against the above-mentioned deficiencies, the GAMLEC partners put forward the following quality standards:

- The physical environment supports the autonomy of nursing home residents, enabling them to move freely in the garden or other facilities such as a restaurant, library or hairdresser; never-ending pathways are available for people with dementia.
- Breakfast, lunch and dinner can be chosen within sufficiently big time slots according to the habits of adult individuals, and meals can also be eaten in

private. Sometimes people do not feel like taking meals in communal areas or they want to enjoy a meal with family members in a private surrounding. A care facility concerned about the well-being of its residents must respond to these wishes without recourse to bureaucratic procedures.

- The independence of the residents in body care and cosmetics is supported, and residents who need help in dressing look well-groomed. Casual clothing can be appropriate and some-times desirable, but may also be a sign that staff is overburdened with their workload. It should be ensured that the residents receive sufficient help and support with more traditional clothing preferences. In general, a well-groomed appearance contributes to self-respect and self-consciousness.

- Assistance in the procurement of cash is provided on request, and purchases of everyday necessities can be done. Spending cash is an important way of feeling capable of functioning in life. If a shop is not within walking distance, purchases should be organized upon request.

- Pursuing individually meaningful activities are investigated when moving into the care facility. The possibility for continuing meaningful activities is very important for the quality of life of care home residents. Reading books, engaging in handicrafts or art activities can be meaningful for older people. Where needed, they must be supported by paid staff or volunteers.

3.4.3 PARTICIPATION

Relations in nursing home are crucially important as it relates to the need of relatedness and belonging. According to Custers et al. (2012), relatedness refers to feeling connected to others or having a sense of belongingness. The most important relations in nursing homes are those between professional caregivers and residents and among the residents itself. "Human closeness with the staff means a lot to me", states an Italian care home resident. "I would like to have more time to spend together just to talk".

Usually scarce time does not allow for paid staff to encourage contacts and relations among the residents. As experience show, residents like to socialize, if they are animated to do so. Mobility restricted or bed-ridden persons, too, can participate in activity offers with adequate technical equipment. However, support is time-consuming and hence often not performed. The same applies even more to events outside the care facilities. The participation in sports, cultural or social events is especially important for the quality of life of people for whom this was a habit in younger years. But:

> In the towns near the nursing home, events take place on weekend evenings, whereas only a few staff members at this time in the nursing home: as a consequence the residents lose opportunities to participate in the social life of the community (Team of care facility, Klaipeda region, Lithuania).

The working conditions of professional caregivers are at odds with the possibility of adequately addressing the social dimension. But participation in social life can also be limited, for example, by creating constraints such as limited visiting hours or lack of space to spend time with family, volunteers and friends.

In order to ensure the social participation of care home residents, the GAMLEC compendium demands the following standards:

- Relatives and other trusted persons are involved in care and invited to work as volunteers. Before moving to a care facility, relatives often care for their family members for a longer period of time. The spatial separation, which is inevitably connected with the relocation, can lead to considerable psychological stress on both sides. This can be alleviated by involving the relatives in nursing and social measures in the facility. Furthermore, relatives are encouraged to actively engage in social opportunities for their family member and other residents.
- Residents are supported in making use of offers in the local environment. The activities of the residents should thus not be limited to what is on offer in the facility. This can be realized by offering visits to cultural or sporting events or by encouraging family and friends to take out the resident to participative events.
- Visits by local people and other guests are encouraged by inviting them into the facility. For example, passing strollers can be encouraged to visit the café, those interested in culture to attend films and concerts, and the local community can be invited to a flea market or an Open Day.
- Contact with circles of friends and acquaintances from times before moving into the nursing home is promoted. Positive effects on the well-being of the residents are ascertained by keeping up continuity. For instance, support can be provided to inform friends and acquaintances about the new address and to invite them personally to events.
- The development of trusting relationships and friendships among the residents is encouraged. In old age, friendships are no longer made as easily as in younger years. Moving into a new living environment such as a retirement home exposes the individual to the risk of personal withdrawal. The facility should offer events in which people can get acquainted with each other and relations between likeminded persons can develop.

3.4.4 HUMAN DIGNITY

Human dignity forms an overarching category that needs to be defined and can cover various topics. From the residents' perspective, as documented by a qualitative analysis from the UK, the most prevalent themes are independence and privacy, followed by comfort and care, individuality, respect, communication, physical appearance and being seen as human. Residents and their families also describe incidents where a resident's dignity had been compromised (Hall, 2014).

Human dignity of residents can be violated when attention is not paid to individual needs and preferences, when there is a lack of listening or when the staff is distracted. Rigid timeframes, financial aspects and a lack of training and awareness lead to a number of deficiencies and flaws. Many temporary personnel and many different staff limit the provision of dignified care because they do not know the persons they are dealing with. The violation of ethical rules and norms of conducts is not necessarily grounded in evil attitudes, as put by one expert, but more likely embedded in the work culture of a care facility.

Grave examples are missing respect in intimate care or fixations of unruly or aggressive persons:

> Human dignity can be violated very quickly, for example, by being touched where you don't want it. Some women don't want men involved their intimate hygiene, but they are told we don't have any other staff, you have to go through that now. Or someone who is restless is fixated in a wheelchair or prevented from getting out of bed, which is of course not allowed.

> *(Head of social services, Maintal, Germany)*

Deprecating or belittling ways of addressing old people have a serious impact on them. "The thing that hurts my dignity most is to be infantilized. A few years ago, my situation was more serious and I needed everything, just like a new-born baby. It is difficult to get used to being cared for" (Care home resident, Bologna, Italy). Persons with dementia or other challenging behaviour are at special risk of being denied their dignity.

There are a number of issues that can also deprive humans of their dignity. For instance, communication can be disrespectful, belittling old people or not taking them seriously: "When communicating with the resident, eye contact must both be at the same level. Confidentiality should be ensured when providing or collecting information. Speak in clear and short sentences" (Vice-manager in care facility, Klaipeda region, Lithuania). As rooms are small or have to be shared with another person, residents can only bring only a few personal belongings when moving to the care facility, depriving them of their personal resources, as illustrated in this statement: "I would like to have the possibility of a single room, to have more space. I had to bring a whole life into one closet" (Care home resident, Bologna, Italy).

Volunteers and relatives, too, must be committed to practices that respect the human dignity of nursing home residents. They could take on functions like empowering residents or being the voice of residents in a case of not enough good care and having no possibility to express it by residents themselves.

With view to the mentioned topics, the GAMLEC compendium proclaims the following quality standards:

- A code of ethics exists that rules the conduct of paid staff and volunteers in situations where interests of the care home residents can be harmed. These guidelines should also be shared with residents, their relatives and other persons of trust.
- The tone of staff and volunteers towards the residents is friendly and respectful, expressing a positive attitude towards the other person and signalling that they like to deal with them. Physical or functional limitations do not constitute a reason to be less respectful of those affected. The tone of staff and volunteers can be regarded as respectful if, despite all possible physical and mental deficits, the other person is acknowledged with all his or her life's achievements.

- Before entering the residents' rooms, staff and volunteers knock on the door and wait for permission to enter. The rooms and apartments are the private retreats of the residents and must be respected as such. Exceptions are permissible in emergency situations or if explicit agreements are made, for example, if persons are hard-of-hearing.
- The residents can furnish their living area according to their own wishes and are supported in doing so. The furnishing of the living environment with personal belongings is elementary for the quality of life of people of all ages and in all health states. Scientific studies have shown that familiar furnishings have a positive effect on people who are suffering from dementia.
- During nursing activities, the privacy of the residents is protected by closed doors, room dividers in double rooms or architectural features, and nursing can be received from persons of the same sex. The latter should not be prevented because of administrative reasons Exceptions are only permitted at night and on weekends if staffing is reduced.
- Procedures concerning the last phase of life and the procedure after death are agreed with the individual resident and documented. In case of ambiguities concerning the organization of the final phase of life and the procedure to be followed after death, relatives or other trusted persons are consulted to comply with the presumed will of the person concerned.

Wikström and Emilsson consolidated their research on the culture of institutionalized care in three emerging themes. Firstly, it revealed an ambivalent mission of care facilities, indicating ambiguity as to whether the nursing home was a place to live in or a place in which to be cared for. The second theme was symbolic power, which encompassed the staff's power embedded in the organization. Finally, an ageist approach to care was observed, which was noted in the way staff considered the residents to be old people who were unable and unwilling to strive for autonomy. The three themes were embedded in the organizational culture and were created and recreated in the interaction between residents and staff (Wikström and Emilsson, 2014).

However, being in need of care does not necessarily impede the realization of quality of life in old age. Very positive comments by German nursing home residents are in contrast with many statements that living in a care facility has negative effects on their quality of life negatively. The breadth of the assessment of identical aspects suggests that this mainly reflects quality differences in performance (Stiehr, 2016). Measured against the needs and desires of older people, the progress towards quality of life in care homes is still in its infancy in all European countries and that there is still considerable potential for improvement.

3.5 SERIOUS BOARD GAME

The GAMLEC game's goal is to promote the quality of life standards on autonomy, participation and human dignity collected in the Compendium. However, knowing the standards of quality of life is not enough to make a significant change in quality of life of dependent seniors. As stated, many solutions that improve quality of life of care home residents require a change in attitude of persons in relation to dependent older people. The game therefore also pursues the goal to enhance general awareness of quality of life, self-reflection, the ability to change perspective and to collaborate, as well as to hone the capacity to empathize with other individuals. In fact, empathy is crucial for bridging the perceived gap between "us" and "them", envisioning dependent older people as "people like you and me", rather than a kind of somehow "different species".

Serious games for adults can offer a valid approach to learning through experience, as well as through reasoning (Connolly, 2012; Hamari, 2014; Odessa, 2013; Ricciardi, 2014). Players are encouraged in a joyful way, and in a safe environment, to reflect on their values, on their actions and the related consequences. This can inspire individuals to reflect on possible alternatives to their habitual behaviour, as well as to find creative solutions for problems.

GAMLEC is a learning game for professional staff in nursing homes, as well as for volunteers and family members of the care home residents. It aims at broadening the view of all players through increased awareness, knowledge and the capacity to cooperate, in order to ensure that standards for quality of life for dependent older people become a tangible reality in everyday life.

3.5.1 BUILDING THE GAME

Based on the assembled concrete examples for autonomy, participation and human dignity, the consortium jointly built cards to play with on the board game. The cards include scenarios for an ideal situation for quality of life and scenarios that present embarrassing situations. Also, the game will include knowledge assessment questions, provide examples of improved built environments or ask the players to give their opinion on a certain topic. The game rules will suggest that players throw dices and that the winner of the game is the one who assembles most quality of life points or is at the finish at first.

The game and cards are supposed to apply as such to all country and all nursing homes in terms of quality standards. This does not mean that they apply to all nursing homes, be it in terms of availability, be it in terms of specific rules or circumstances of the single nursing home. The game will not make "requirements" but offer inputs of what exists, of best practices, and invites people to creatively think what is applicable in their specific situation. The game is supposed to offer inputs to thinking of what possibly can be done to improve quality of life in any given situation, step by step.

The consortium will jointly define an overall general standard for the game in English but will consider cultural differences and ethical issues. Cultural differences, such as alcohol consumption, sexuality in old age and in the nursing homes, sexual diversity cards that are not acceptable in all partner countries will not be translated or offered in these countries, but only remain part in the English edition of the game. On the other hand, topics that are culturally/legally accepted in one specific country can be added there, such as euthanasia in the Netherlands. Besides that, the board game will offer online flexibility to players to create and add their own playing cards that fit in their local context.

3.6   CONCLUSIONS

Quality of Life is, as stated in the introduction, about the well-being of individuals. It is subjective and defined differently, depending on the individual's cultural background and individual preferences. Nevertheless, above and beyond all individual differences, all people strive for joy in their life, independent from age, gender, collective or individual culture, and, very important, independent from age or occurring functional disabilities. The GAMLEC board game therefore offers the proven methodology of serious gaming to open up the discussion on quality of life in an easy accessible way and have fun to play. The impacts of the game will be evaluated within the frame of the project in 2021.

Limitations of the study are that it is based on the desk research and interviews of experts in four European countries only. The impacts of the game in practice have not been evaluated yet. The GAMLEC board game will become freely available for each individual or organization. Further research to measure the effects of the game on knowledge, values and opinions on quality of life in nursing homes is most welcome.

REFERENCES

Connolly, T. M. et al. (2012, September). A systematic literature review of empirical evidence on computer games and serious games. *Computers & Education*, 59(2), 661–686, Doi: 10.1016/j.compedu.2012.03.004.

Custers, A. et al. (2012). Relatedness, autonomy, and competence in the caring relationship: The perspective of nursing home residents. *Journal of Aging Studies*, 26, 319–326.

Evans, E. A. (2018, February). Care home manager attitudes to balancing risk and autonomy for residents with dementia. *Aging and Mental Health*, 22(2), 261–269. [Online] Available from https://www.ncbi.nlm.nih.gov/pubmed/27768393 [accessed June 13, 2020].

European Commission, Lithuania. (2016, October). *Health Care & Long-Term Care Systems.* [Online] Available from https://ec.europa.eu/info/sites/info/files/file_import/joint-report_lt_en_2.pdf, [accessed July 8, 2020].

European Social Policy Network. (2018). *Challenges in Long-Term Care in Europe. A Study of National Policies.* Brussels: Country reports of Germany.

Eurostat, Ageing Europe. (2019). Looking at the lives of older people in the EU, 2019 Edition. Doi: 10.2785/811048.

GAMLEC - Game for Mutual Learning in Elderly Care, Erasmus+ project GA 2019-1-DE02-KA204–006492. (2020). *Compendium on Criteria for the Quality of Life of Care Home Residents and National Reports for Germany.* Italy, Lithuania, The Netherlands. www.gamlec.eu.

Hamari, J., Koivisto J. and Sarsa H. (2014). Does gamification work? -- A literature review of empirical studies on gamification, 2014. *47th Hawaii International Conference on System Sciences*, Waikoloa, HI, 3025–3034, Doi: 10.1109/HICSS.2014.377.

Hedman, M. et al. (2017). Caring in nursing homes to promote autonomy and participation. [Online] Available from https://journals.sagepub.com/doi/abs/10.1177/09697330177036 98?journalCode=neja#abstract [accessed 13.06.2020].

HHM. (2015). Leidraad verantwoorde personeelssamenstelling voor verpleeghuizen in Nederland.

Hall, S., et al. (2014). Patterns of dignity-related distress at the end of life: a cross-sectional study of patients with advanced cancer and care home residents. Palliative Medicine, 1118–1127. Doi: 10.1177/0269216314533740.

Klein, B. and Gaugisch, P. (2005). *Gute Arbeitsgestaltung in der Altenpflege, Selbstbewertung als partizipationsorientierte und nachhaltige Methode für die gesundheits-förderliche Arbeitsgestaltung in der Pflege, im Rahmen der Initiative Neue Qualität der Arbeit (INQA), Bundesanstalt für Arbeitsschutz und Arbeitsmedizin (Hg.)*, Dortmund.

Kratzer, M. (2011). *Lebensqualität im hohen und höchsten Alter im städtischen Bereich. Eine Unter-suchung von 75- bis 95-jährigen Wiener und Wienerinnen in Privathaushalten und Altersheimen*, Dissertation. Universität Wien, Wien. [Online] Available from http://othes.univie.ac.at/18629/1/2011-10-11_0348574.pdf [accessed November 25, 2019].

Kuratorium Deutsche Altershilfe (KDA). *Vom Pfegeheim zur Hausgemeinschaft*. [Online] http://www.infaqt.de/media/files/nrw_vom_pflegeheim_zur_hausgemeinschaft.pdf [accessed June 17, 2020].

Kwaliteitskader Verpleeghuiszorg. Published by Zorginstituut Nederland, 2017. [Online] https://www.zorginstituutnederland.nl/publicaties/publicatie/2017/01/13/kwaliteits-kader-verpleeghuiszorg [accessed October 25, 2020].

Odessa, J. et al. (2013, December). Developing the serious games potential in nursing education. *Nursing Education Today*, 33(12), 1569–1575. Doi: 10.1016/j.nedt.2012.12.014.

Ricciardi, F. and Tommaso de Paolis, L. (2014). A comprehensive review of serious games in health professions. *International Journal of Computer Games Technology*, 2014, Article ID 787968, 11 p, Doi: 10.1155/2014/787968.

Stiehr, K. et al. (2016). *Lebensqualität im Alter: Kriterien für eine zielgruppengerechte Verbraucher-information. Bericht an das Ministerium der Justiz und für Verbraucherschutz*, Berlin. [Online] Available from https://www.heimverzeichnis.de/fileadmin/Basismaterial/Heime/BerichtLebensqualitaetImAlter.pdf [accessed June 11, 2020].

Theobald, H., Szebehely, M. and Preuß, M. (2013). Arbeitsbedingungen in der Alten-pflege. Die Kontinuität der Berufsverläufe – ein deutsch-schwedischer Vergleich.

Wiese, C. (2015). *Einflussfaktoren von Lebensqualität im Alter: eine systematische Übersichtsarbeit*, Dissertation, Staats- und Universitätsbibliothek Hamburg, Hamburg.

Wikström, M. and Emilsson, U. M. (2014). Autonomy and control in everyday life in care of older people in nursing homes. *Journal of Housing for the Elderly*, 28(1), 41–62. [Online] Available from https://www.tandfonline.com/doi/abs/10.1080/02763893.2013.858092 [accessed June 13, 2020].

World Health Organization / Milberg Memorial Fund. (2000). *Towards an International Consensus on Policy for Long-Term Care of the Ageing*, https://www.who.int/ageing/publications/long_term_care/en/ [accessed July 07, 2020].

World Health Organization. (2015). *World Report on Ageing and Health*, https://apps.who.int/iris/bitstream/handle/10665/186463/9789240694811_eng.pdf?sequence=1&ua=1 [accessed May 15, 2020].

4 International Students' Well-Being Achievements through the Lens of the Capability Approach
A Case Study

Icy Fresno Anabo, Iciar Elexpuru-Albizuri,
and Lourdes Villardón-Gallego
University of Deusto

CONTENTS

4.1 INTRODUCTION

Although governments have long been concerned with improving life standards, promoting quality of life and well-being has particularly gained a strong interest in the last few decades. The evolution of related empirical and theoretical work over the years shows how quality of life has come to be understood as a multidimensional construct embodying both objective (such as health, education, employment, and participation outcomes) and subjective components (such as happiness, life satisfaction, and sense of meaning) (Eurofound 2017; Eurostat 2015).

In this context of growing academic scholarship and policy initiatives on the topic, there is a wide consensus on the centrality of education as an instrument for living a good life. It is viewed to contribute to positive health outcomes, with educated parents being more likely to make informed decisions that reduce child and maternal mortality rates (UNESCO 2016). Completing tertiary-level education has also been shown to improve labour market outcomes (such as resilience to long-term unemployment and stronger earning capacity over time) (OECD 2019) and improved political interest (Easterbrook, Kuppens, and Manstead 2016). Subjectively, educated individuals are more likely to report being happy and satisfied with their lives (Eurostat 2015), with evidence demonstrating higher levels of education as instrumental in promoting personal relationships, vitality, engagement, and resilience (Jongbloed 2018).

The goal of this chapter is to contribute to the growing body of literature on the impact of educational experiences on students' quality of life and well-being using a qualitative and capability-based analysis of students' narrated experiences. Beyond formal qualifications and the use of quantifiable indicators, narratives can be a robust instrument to understand and evaluate people's inner lives (Bauer, McAdams, and Pals 2006). Additionally, using the Capability Approach's (hereon referred to as CA) broad conception of well-being as freedom constitutive of individual and social elements can provide a framework that amply captures what it means to live well. Through this analysis, we seek to illuminate the well-being achievements of students in and through education and provide a more granular account of how an educational experience can mediate such outcomes.

4.2   THE CAPABILITY APPROACH

The Capability Approach originally proposed by Amartya Sen relies on Aristotle's concept of human flourishing as the ultimate goal of societal arrangements. It is a normative framework in the sense that it provides claims about how we conceptualize and evaluate societal progress (Robeyns 2017) whereby the focus of political and social interventions lies on people's *capability*, or their effective freedom to pursue beings and doings that they have reason to value (Sen 1992, 1999). He differentiates between *capability* – which refers to the range of an individual's possible options of beings and doings – and *functionings*, which refer to actual achieved states. This is a subtle yet important distinction, since it frames well-being interventions not as prescriptions but as actions to enable individuals to make reasoned choices and to guarantee that effective options are available (Walker 2008). One of CA's most prominent contributions is its expansive view on well-being, providing a robust framework around which public policies and interventions may be designed.

The CA's normative proposal has since then been used to examine the various dimensions of the higher education endeavour (Robeyns 2017) which, for many scholars, serve not just economic but also personal and social functions (Boni and Walker 2013; Brighouse and Unterhalter 2010). Nussbaum's work (2000, 2011) is an important reference in this domain. Contrary to Sen's reluctance to endorse exact configurations of what individuals' capability set should look like, she argues that certain fundamental freedoms must be stipulated as thresholds or minimum guarantees for a truly dignified life. This supports Saito's (2003) claim that freedom may be used ill-naturedly; hence, there is a need to evaluate and promote those that are supportive of human flourishing. Nussbaum (2011) outlines these specific forms of freedom in her list of ten central capabilities: life; bodily health; bodily integrity; senses, imagination, and thought; emotions; practical reason; affiliation; other species; play; and control over one's environment through political participation, possession of property, and work. For her, achieving these fundamental freedoms not only supports and ensures human dignity but also enhances a myriad of complex and intrinsically important states and activities that allow individuals to live truly enriching lives. Scholars have demonstrated that the use of lists like that of Nussbaum may be fruitful in the field of education, especially when making decisions regarding teaching approaches and pedagogies (Walker 2006) and defining education's desired outcomes (Flores-Crespo 2007; Young 2009). By embodying both individual and social aspects, Nussbaum's work is able to incorporate a social dimension that resonates with higher education's plural functions.

This qualitative analysis draws on the CA and Nussbaum's list to examine capability achievements as an expression of well-being through the narratives of a group of students who completed an international master's programme in education studies. Specific research questions include:

1. Which capabilities did students achieve in relation to their international study experiences?
2. Which features of their international study experience shaped their achievements, and how?

The projected contributions of this empirical study are twofold: to identify international education's contributions to well-being and to illustrate the usefulness of the CA as an evaluative framework.

4.3   PARTICIPANTS AND METHODS

The study adopts an embedded case study design (Yin 2018) that considers a mobility programme in education studies as the main unit of analysis and the participants as subcases. The chosen programme was financed by the European Commission within the Erasmus Mundus (EM) scheme for joint master's degrees, which was composed of an international consortium of three European universities, one non-European university, and one non-university institution.

We selected the research participants through a combination of purposive and convenience sampling (Saumure and Given 2008). We determined the initial list of candidates by considering the inclusion of a variety of students based on their graduation year, country of origin, gender, and professional background. Based on these criteria, 52 eligible participants were contacted through email or messaging platforms, 22 of whom confirmed and followed through with the interview. A summary of their demographic information is shown in Table 4.1. The participants consisted of 14 females and 8 males. Half of the total number of participants ($n = 11$) graduated within the last 5 years, while the other half ($n = 11$) within the last 6–10 years. The participants represent nine regional categories identified in the list published by the European Commission (2019).

We sent a digital consent form to the participants along with an online questionnaire collecting their demographic data. We then arranged semi-structured online interviews lasting between one to two and a half hours between May 2018

TABLE 4.1

Participants' Profiles

Pseudonym	Cohort Group[a]	Sex	Regional Category of Participant's Home Country[b]
Joseph	1	M	Asia
Julia	1	F	Russian Federation
Ana	1	F	Eastern Partnership Countries
Ivan	1	M	Asia
Jane	1	F	Other Industrialized Countries
Katrina	1	F	Eastern Partnership Countries
Lei	1	F	Asia
Maja	1	F	Non-EU Programme Country
Lily	1	F	Asia
Omar	1	M	Asia
Catarina	1	F	Latin America
Ayo	2	M	Africa, Caribbean and Pacific Countries
Adrian	2	M	Member States of the European Union
Dara	2	F	Asia
Li	2	M	Asia
Fernanda	2	F	Latin America
Sophia	2	F	Non-EU Programme Country
Jack	2	M	Member States of the European Union
Dao	2	F	Asia
Victoria	2	F	Central Asia
Martin	2	M	Member States of the European Union
Mariam	2	F	Eastern Partnership Countries

[a] Group 1 refers to graduates of the first five cohorts (2006–2010) and Group 2 to those from the latter five cohorts (2011–2015).

[b] Based on Erasmus+ Regional Categories (European Commission 2019).

and January 2019. We interviewed each participant once and, when necessary, sent post-interview correspondence to clarify specific aspects of the data. The responses were recorded, stored in a password-protected device and cloud service account, and transcribed. General (questions 1 and 2) and specific (questions 3 and 4) prompts were used to evoke participants' capabilities:

1. Can you tell me about your significant experiences during the programme?
2. What were the highs and the lows during this time?
3. What were the contributions of the programme to your life?
4. What new opportunities did you have resulting from your participation in the programme?

The main interest lay in finding patterns in the interview transcripts along two strands: the participants' achieved capabilities and the contributing factors that led to them. Using the directed content analysis approach (Hsieh and Shannon 2005), the data underwent iterative rounds of systematic coding. Provisional coding was used for the exploratory phase (Saldaña 2013) by using categories and concepts from existing sources. Nussbaum's ten central capabilities list was used as main codes to tag the participants' capability achievements, four of which emerged as the most salient themes in the narratives analysed, as summarized in Table 4.2. Meanwhile, broad concepts based on critical realist social analysis (Danermark et al. 2002) such as 'structure', 'agency', and 'programme-related' were employed to identify excerpts signalling possible contributing factors.

TABLE 4.2
Coding Guide

Main Codes	Sub-codes	Sample Excerpts
Practical reason	Visualize life plans	'It's broadened my vision so that I became more aware of what actual functions can be attached to that specific academic programme'.
	Form conceptions of a good life	'I learned to live with less and I realised that I don't need so much to be happy'.
	Make critical and socially responsible choices	'I think this attitude to nature, like being more conscious about the fact that you're part of the world so you have to protect it...it was always a part of me... but it became very strong when I saw like-minded people'.
Affiliation	Social belonging	'The experience made me a little bit more autonomous in being cosmopolitan. It also inspired in me this idea of being part of the European project'.
	Professional and alumni networks	'I needed some education resources for my work, so I joined the Alumni Association and I found the resources'.
	Social relations (friendships and romantic relationships)	'I gained best friends and friends for life'.
Senses, imagination and thought	Pleasurable experiences	'That is the place where for the first time in my life I felt that I had a really strong affinity and liking for art, culture, and music'.
	Critical knowledge of self and the world	'It made me question so much of what I had previously taken for granted'.
	Field-specific, intercultural and language skills and knowledge	'Going to the programme provided me with the tools that I am using now in my professional practice, especially doing comparative analysis'.
	Confidence and resilience	'I suddenly got invited from different other departments from within the university... so taking that leading role was a very empowering feeling'. 'The travel itself ...gave me more resilience and helped me to embrace change as it comes'.
Work	Participate in work	'I got involved with the Erasmus Mundus Association, and through that affiliation people learned about some of my work and opportunities came up'.

The next round of coding involved subcoding (to reach a more detailed analysis of capability achievements) and causation coding (to attribute and connect certain capabilities to explanatory factors) (Saldaña 2013). Subcodes for capability achievements were taken from definitions provided by Nussbaum (2011) herself and other CA scholars. For example, the capability 'senses, imagination, and thought' would include using one's imagination, having the freedom of religious exercise, or engaging in pleasurable or aesthetic experiences.

Drawing causal relationships relied on a combination of two approaches: using participants' own perceptions on what shaped their outcomes and employing abduction and retroduction. Abductive reasoning involves engaging with data to form a plausible hypothesis explaining what was observed, while retroductive reasoning relies on a non-formalized inferencing technique relying on a reconstruction of the conditions that may explain the phenomenon in question (Danermark et al. 2002). In this study, abduction and retroduction were based on students' perceptions on what shaped their outcomes, a review of the existing literature, and comparisons between student sub-cases. This was followed by a deliberation phase to clarify and refine the concepts and the check for the appropriateness of codes in the tagged excerpts. A final round of coding ensued to reorder and finalize the relationships between the concepts, sketching the unique combinations of factors that led to specific capability outcomes.

To anticipate and address the possible ethical concerns arising from the project, a work plan for data collection and reporting methods was drafted and approved by the Academic Committee of the relevant faculty. Several guidelines for conducting online research with human subjects were also reviewed (Anabo, Elexpuru, and Villardón-Gallego 2019).

4.4   FINDINGS AND DISCUSSION

This section illustrates the extent to which participants achieved various forms of capabilities, which we equate to the achievement of well-being. These outcomes allude to Nussbaum's practical reason; affiliation; senses, imagination, and thought; and work capabilities. Our results also show that overall, these well-being outcomes were shaped by the international aspects of the programme as well as a range of non-international, agentic, and contextual elements.

4.4.1   PRACTICAL REASON

Practical reason concerns 'being able to form a conception of the good and to engage in critical reflection about the planning of one's life' (Nussbaum 2011, 34). It suggests a sense of critical autonomy (Walker 2006) and manifests as being able to visualize or plan one's life (Flores-Crespo 2007; Walker 2007), form one's own conceptions about what a good life looks like (Walker 2007), and make critical and socially responsible choices (Walker 2006).

In terms of visualizing life plans, Martin and Julia found their participation in the programme to be clarifying and expansive with regard to their professional choices post-graduation. According to Martin, its wide and international array of theoretical and practical research exercises (curricular content) 'broadened his vision' and made him discover the kinds of positions that he could pursue afterwards. Similarly, Julia, who had previously planned to work as a translator, shared that her internship experience during the programme widened her career choices. Meeting people she admired and developing a sense of confidence about the value she could provide in the profession led her to appreciate the discipline and consider it as a career path. She shares:

> When I was doing my (translation) studies, education was considered a lower option... but working here for my internship, I met very bright people so somehow you see you can take on projects, you can move things.

While participating in the programme allowed Martin and Julia to define their professional plans, Dao struggled to find clarity. Her own reflection during the interview led her to conclude that her own lack of foresight kept her from having a more concrete vision of the future. She shares:

> Back then, I didn't even imagine what my life would be after that. I just had this illusion... I just wanna live abroad to see what life will bring.

Dao's experience is not uncommon. Hemmer et al.'s (2011) study suggests that international (especially non-EU) students tend to have broader future goals at the beginning, which would then require a more focused effort from staff to provide career planning opportunities. Another participant, Maja, also alluded to the importance of professional orientation and suggested that this may be supported in the programme by having dedicated mentors. Relevant studies support this view, with mentoring being one of the features that a large number of students felt was lacking in EM (ICU.net AG 2014; Kruger, Klein, Pinkas et al. 2017).

The second dimension of practical reason is forming a conception of a good life. For Fernanda, her participation was instrumental in its achievement to the extent that the mobility offered allowed her to live abroad. As a result, this expanded her notion of a meaningful life and shifted her values towards simplicity.

> I learned to live with less... It changed my values in terms of what I need to live and I realised that I don't need much.

Lastly, social responsibility was an enhanced dimension among the participants, which coincides with Tarrant, Rubin, and Stoner's (2014) findings on the mediating role of international education in this regard. Victoria, for instance, shared that she became more aware of her environmental impact that led to a value shift towards making ethical choices. She related this achievement to the diverse kinds of people and perspectives she encountered in such a mobility programme.

> My values completely changed. For instance, I had a friend and she's all like, sustainable development, recycling, and all these European values. We don't even recycle back home... So I think this attitude to nature, like being more conscious about the fact that you're part of the world so you have to protect it... I think it was always a part of me... but it became very strong when I saw like-minded people.

This sense of others-orientedness, which spills over as a collective benefit of living well together (Deneulin and McGregor 2010), was also developed by Omar through the programme. He reports that he now engages more critically with the way he behaves and interacts with others, which he attributed to his interactions with his peers and supervisors. He shares:

> Older things got connected with new ways and understandings of life, like being rule-abiding, being very respectful to others while communicating, being physically more controlled while interacting with others, being more critical but at the same time learning to accommodate others' opinions, and thinking in terms of how other people's opinions matter and why.

In the narratives provided above, it seems that the achievement of the practical reason capability such as visualizing life plans, forming a conception of a good life, and developing social responsibility are linked to another central capability: affiliation. Clearer plans for the future appear to unfold when students feel inspired by fellow co-workers or receive valuable guidance from mentors. Quality of life appears to be linked to positive relationships and friendships. Environmental awareness and socially conscious behaviours, which both contribute to collective well-being, are likely to manifest when these values are modelled by and shared with peers. These findings support Nussbaum's claim that the practical reason and affiliation capabilities are mutually enriching (Wilson-Strydom and Walker 2015).

4.4.2 AFFILIATION

The capability of affiliation refers to 'being able to live with and toward others' (Nussbaum 2011, 34) and alludes to the following dimensions: social belonging (Narayan and Petesch 2002), social relations, and social networks (Robeyns 2003; Walker 2006, 2007). Engaging in international programmes can be particularly enriching in this regard as students stretch and reconfigure their notions of affinity.

The first dimension of this capability – social belonging – was reportedly achieved by Katrina and Maja. Katrina alluded to a sense of membership to a vibrant academic community made up of excellent practitioners and intellectuals in the field. Her narrative suggests that the primary mediating factor was the academic environment itself instead of the fact that it was an international programme. Meanwhile, another participant, Maja, specified the physical mobility component as a factor that positively reconfigured her cultural identity as being cosmopolitan and European. She shares:

> It was an absolutely crucial thing that I went out of the region I grew up in. So basically the experience kind of made me a little bit more autonomous in being cosmopolitan. It also inspired in me this idea of being part of the European project.

Maja's narrative runs parallel with existing evidence on the correlation between European-type mobility and the students' development of European identity (Mazzoni et al. 2018; Mitchell 2012, 2015). EM survey findings also point out that developing a positive attitude towards the EU has always been a noteworthy outcome of EM through the years, especially among non-EU citizens (Terzieva and Unger 2019).

The second affiliation capability, social relations, was also reportedly achieved by many participants through the development of friendships and romantic relationships.

For Ivan, experiencing close friendships has been one of the most valuable contributions of the programme both from a learning and social standpoint:

> It's a very collegial and stimulating experience with them (classmates). It was not just the academic side of things, we also traveled together... I gained best friends and friends for life.

Moreover, several participants were able to meet their partners during the programme. This is the case for Victoria, who felt that her changed values prompted her to seek a relationship in the host country:

> I kind of felt that with these changed values, I need someone like-minded ... someone with European values. I knew that back home, I won't be able to talk to men in the same way.

Professionally, the affiliation capability manifested as being part of professional and alumni networks. This is the case for Li, who found that it helped him access valuable resources as part of international student groups. Similarly, Ayo viewed the international professional network he gained in the programme as a means to mediate with his immediate community. His participation allowed him to meet international people and maintain a professional network for cross-country cooperation. As a result, he was able to have a voice in the shaping of international policy statements related to his field.

4.4.3 Senses, Imagination, and Thought

Nussbaum (2011) defines this capability as 'being able to use the senses, to imagine, think, and reason – and to do these things in a "truly human way"' (p. 33). It alludes to freedom of expression and engagement in pleasurable experiences of one's own choosing (e.g. music and literature). For Flores-Crespo (2007), it also constitutes developing knowledge, abilities, confidence, and self-reliance. Overall, the senses, imagination, and thought capability can serve both aesthetic and instrumental purposes.

Several participants achieved the freedom to engage in pleasurable experiences as a result of their participation, including Joseph and Katrina who both were in their mid-careers when they decided to join the programme. Joseph considered it as the 'right kind of break' he needed in his career, allowing him to fully immerse in learning and experience different cultures. Katrina describes the value of this unique opportunity to return to full-time studies:

> I found out that in professional life you have to get things done. Not that you don't have to do the same in academia, but the academic environment allows you sufficient time and gives you enough resources to pursue what you consider to be the truth. In professional life, that's a luxury.

This kind of immersive experience is made possible by the fact that most EM programmes are designed to be completed full-time, with most students benefiting from full scholarships. The EM scheme offers more than 1,000 scholarship places every year (European Commission n. d.), allowing many the chance to focus entirely on their studies.

Another participant, Omar, appreciated the opportunity to engage in cultural experiences during his time in one of the host countries. Coming from a developing country, he points out the rich aesthetic experience as mediated by experiencing life abroad:

> I loved the language, the culture happening around me where elderly people had a strong vibrant social life… That is the place where for the first time in my life I felt that I had a really strong affinity and liking for art, culture, and music. I didn't have that back home… every day was like a war with yourself and others.

The programme's impact was also linked to a more informed and critical knowledge of one's self and the world. Often, when some distance exists between the person and his or her familiar context through arrangements such as physical mobility, individuals can 'become aware of a certain relativity of one's functioning' that is both destabilizing and enlightening (Cicchelli 2013, 206). This is true for Ivan, who alludes to the physical mobility of living in another country that triggered a process of reflection.

> Being away from your family, your friends, and even from your small community allows you that degree of freedom to reflect, to assess who you are and what you are, and what you're capable and not capable of doing.

Meanwhile, for Lei, the combined effects of classroom diversity, academic content, and the critical pedagogy adopted were particularly significant in the deep-seated shifts she experienced. Jack singled out the critical pedagogy employed in their classroom interactions as crucial for this achievement:

> The programme was about policy, so I was thinking of the economic context, political context of different countries, how all the societies organise themselves, their different political systems and how they impact their education systems… and we had students from Eastern Europe, Africa, India, and Pakistan, learning about democracy and debating every day… that's something I had never known in my country.

(Lei)

> I was stretched academically throughout the degree, especially during the first year in one of the host universities where they really pushed us to think critically about everything. It made me question so much of what I had previously taken for granted, and allowed me to develop more thorough reasons for the ideas and opinions I advocate for.

(Jack)

These findings coincide with Cai and Sankaran's (2015) findings on the combined effects of carefully designed coursework and student mobility. In their research, students developed critical thinking through thought-provoking questions, exposure to different disciplinary contexts, and the comparisons they make while being immersed in a different culture.

Another dimension of the senses, imagination, and thought capability involves knowledge and skills acquisition. For students like Dara, Ivan, Adrian, and Catarina, these gains were instrumental to their professional practice later on. Dara appreciated acquiring field-specific knowledge such as theories, paradigms, and methodologies through the programme's diverse and international curricular content. A similar outcome was reported by Ivan, who shared how the multi-site mobility embedded in the programme contributed to his professional practice. Through his exposure to unique academic traditions and cultures, he was able to enhance his comparative analytical sense. This corroborates Bracht et al.'s (2006) findings that sociology-related disciplines such as the case presented in this study particularly benefit from cross-border study arrangements since effective social analyses are sharpened through comparative and international perspectives. Additionally, participants also alluded to the linguistic and cultural gains brought about by the programme's pedagogy (through group work with students from diverse backgrounds) as well as the use of English as a medium of instruction. The former, for example, developed Adrian's intercultural skills, which enhanced both his personal and professional life. Meanwhile, Catarina referred to the latter as a mediating factor in her pursuit of an international career due to the oral and written proficiency she gained in the programme.

Lastly, Omar and Jack reported gains in professional confidence. For Omar, becoming more professionally competent and acquiring international qualifications allowed him to take a leading role at work and feel empowered as a result:

> I suddenly got invited from different other departments from within the university to design whole courses and conduct them, which they used to hire international faculties for. Taking that leading role was a very empowering feeling and leading it successfully was rewarding at the same time.

Beyond the professional realm, a sense of personal confidence and resilience was also a positive outcome for Jack. He points out that having the chance to live and settle in different cities contributed to this achievement:

> The travel itself helped me to grow as constantly having to adapt to a new country, a new education system, and new ways of doing things gave me more resilience and helped me to embrace change as it comes.

4.4.4 WORK

The previous section on senses, imagination, and thought illustrates the idea that it reinforces another relevant capability – that of work. While it is not specified as a central capability as such in Nussbaum's (2011, 34) list, she refers to work as a dimension of the capability of having control of one's own environment, more specifically as 'being able to work as a human being, exercising practical reason and entering into meaningful relationships of mutual recognition with other workers'. This definition shows that work is a complex capability involving a combination of other fundamental freedoms such as practical reason and affiliation. The narratives that follow illustrate the relationships between these capabilities, offering a nuanced picture of students' various work outcomes as mediated by agentic, contextual, and programme-related factors.

Several participants went on to pursue promising careers either immediately or several months after degree completion. This positive outcome appeared to be influenced by individual factors (personal choice regarding job and place of residence as well as previous work experience) and a combination of programme and contextual factors (such as the symbolic value of a foreign and/or field-specific qualification). The former is illustrated by the fact that those who had jobs waiting back home and chose to go back such as Lily and Katrina experienced relatively smooth transitions – they were able to apply their new knowledge and assume more advanced roles in their jobs. Meanwhile, the latter point is supported by both Martin's and Ayo's experience. Martin attributed his positive work outcomes to the prestige of the universities he attended while Ayo shared that having a degree in education studies allowed him to shift from being a linguist to his valued career of working in the education sector. He shares:

> In here, we don't get to choose what we study, especially at bachelor's level … it's done centrally by the Ministry of Education. My bachelor's was in English language, that's why I studied Linguistics for my master's. Although I was good at it, I didn't want to be doing it for the rest of my life because I live in a society with several problems. I didn't have the solution if I studied linguistics.

In addition to the factors mentioned above, the participants' narratives also reveal that the achievement of work capability is shaped by gains in other capabilities such as practical reason, affiliation, and senses, imagination, and thought. For example, having clearer life plans (practical reason) combined with the acquisition of specific personal and professional knowledge and skills (senses, imagination, and thought) allowed Martin to channel his attention to specific paths and make more strategic investments in time and effort:

> We studied sociology, learning theories, policies and without a strong focus, I don't think I could have managed. What I see is that many of my peers who were succeeding in the programme had a strong vision beforehand.

Meanwhile, participants' international professional networks were also instrumental to their positive work outcomes. For example, Ayo was able to tap on the network of international contacts he made during his internship experience for the subsequent community projects he spearheaded in his work at a university in his home country. Jane also shared how being part of the alumni association opened work opportunities for her and served as a crucial break in her work life. Aside from policy-oriented careers, many also pursued the academic track by working as lecturers or pursuing Ph.D. studies. For Maja, maintaining contact with a renowned scholar in the field as her master's thesis supervisor allowed her to secure her recommendation for a Ph.D. grant.

In spite of these benefits, other participants experienced challenging transitions back to the labour market. In Dao's case, her personal decision to stay in the host country to pursue a new relationship posed work constraints related to navigating a foreign labour market. She shares:

> I didn't realise that I would meet my boyfriend in the third semester and it would work out that I decided to continue this relationship and wanted to stay… I needed to survive so I took any job I could.

Similarly, the cases of Ivan and Ana illustrate how personal decisions as a function of agency led to employment issues. Given their personally motivated decision to stay in the host country post-graduation, they had to resort to either unqualified or undesired jobs in order to stay employed. They attributed these outcomes to the temporal-contextual factor of timing, having been thrust into the labour market during the economic crisis in Europe. Additionally, Sophia – who also shared the same difficulties in landing a job in the host country – alluded to the combined contextual effects on the one hand (e.g. her lack of local language competency and biases against foreigners in the host country's hiring practices) and programme-related factors on the other (programme's weak signalling power to employers) as contributory to her difficulties in finding a relevant job. Fernanda also alluded to the last point saying:

> Given the name of the course, I thought it would open me doors because it's a wide concept. It's too wide and I realised that I'm not a specialist in anything... in the end you don't have specific skills that you can write you can do.

Mariam also alluded to the mismatch between the programme's signalling effect and employer expectations. Although she was willing to wait for the 'right' job to come along, she found it confusing that her degree did not have the desired effect on employers:

> We obviously know what it has given us and how it helped us develop professionally, but this experience of job hunting gave me the impression that it is not as well recognised as it should be... they are just asking for a quick, short course in management or something like that.

Challenges also had to be managed by returning students. Ivan, who went back home to work in education policy and research, mentioned his potential employers' interest yet lack of familiarity with the scope of his degree, which was largely focused on the European system. In the end, this lack of awareness from the demand side proved to be a huge barrier.

Given these contextual factors, Jane alluded to a programme-related measure that could have mitigated this risk. She thought that participating in an internship could have been a good experience in order 'to connect more with the workforce, people, and the community'. Her view supports the key role of work-based learning arrangements in preparing students for work and aligns with other related studies' findings on the untapped potential of this mode of learning in European mobility programmes (Hemmer et al. 2011; ICU.net AG 2014; Kruger, Klein, Pinkas et al. 2017; Terzieva and Unger 2019).

4.4.5 The Programme's International Dimensions as Mediating Factors

As depicted in Figure 4.1, the participants' narratives illustrate the extent to which the programme's international components facilitated their capability achievements. For instance, in the realm of practical reason, international dimensions such as mobility and cross-cultural topics in the curriculum notably mediated the formation of conceptions of a good life and social responsibility insofar as it expanded the participants' spheres of experience and affiliations. Through physical mobility and exposure to a variety of perspectives, the participants were able to generate comparisons and thus engage in reflections about their own values. On the other hand, having clearer life plans was contingent on the features of the programme itself, including field-specific academic content, practical components of the coursework, and the people that students interacted with in the academic environment.

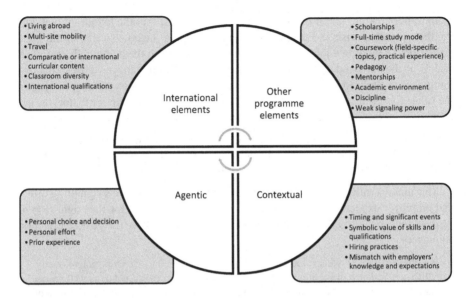

FIGURE 4.1   Factors that shaped participants' capability achievements.

Meanwhile, the dimensions of the affiliation and senses, imagination, and thought capabilities were shaped by the international features of the programme such as physical mobility, diversity, and international curricular content. They were able to widen the spectrum of participants' socio-cultural exposure, allowing them to gain significant aesthetic and instrumental learning experiences. Lastly, their work capability showed mixed results, with agentic and contextual elements playing a modulating effect for many of the participants' post-graduation trajectories.

4.5   CONCLUSIONS

The analysis presented aims to identify the participants' capability and well-being achievements through their participation in an international education programme and illuminate the features of their experience that led to such outcomes. It was shown that the international case analysed was instrumental to the achievement of fundamental freedoms such as practical reason, affiliation, senses, imagination, and thought, and work, which illustrate the personal and social dimensions of such an educational arrangement. While these capabilities were treated as separate concepts in the previous section, they are inherently interconnected, with certain capability combinations being mutually reinforcing or triggering unwanted constraints. It was also found that the participants' capability achievements were a product not only of the international dimensions of the programme (living abroad, multi-site mobility, and classroom diversity) but also of other non-international programme elements (such as the programme's discipline, the mode of study offered, the availability of work-based and practical learning arrangements, and the symbolic value of their degree), individual agency (personal choice, effort, or previous experiences) and contextual factors (such as timing and significant events, hiring practices, and employer expectations).

This analysis presents a number of insights for policy and practice. Firstly, it illustrates how the value of international education may be perceived, including its potential to organically trigger critical reflection, significantly shift students' values and life trajectories, and allow them to live enriching and stimulating life experiences. These are valuable outcomes in themselves not only because they can lead to instrumentally beneficial results but also because they support the intrinsic value of students' lives and choices through education. Secondly, it offers insights into the limits of international forms of study and shows certain gaps to be addressed. Notably, their potential in facilitating the achievement of work capability cannot be fulfilled by student mobility alone. Although a foreign degree is often perceived as a ticket to a good job, the individual and contextual factors at play are equally crucial. As such, programmes could intervene by strengthening the agentic capacities of students for making strategic and informed choices through career planning and insight, work-based learning opportunities, and close mentorships.

4.6 LIMITATIONS AND FUTURE RESEARCH

Moving forward, future studies are encouraged to build upon the inferences derived from this analysis through quantitative methods and further explore the scope of the CA's applicability in how education more broadly mediates collective well-being and social justice. For instance, it would be worthwhile to closely examine the distributive element of educational choice, participation, and success in international studies along gender, economic, and socio-cultural lines. It is worth noting that the participants came from rather advantaged backgrounds with high levels of existing capital. As such, it will be interesting to study the types of pre-existing capabilities that allow people to access this kind of learning arrangement and possible pathways to address systemic imbalances. Additionally, the study has invoked the social dimension of the CA by highlighting the social and political conditioning of capability achievement and analysing personal capabilities that hold social benefits. A further step would involve a deeper analysis of how education – specifically international higher education – can mediate Evans' (2002) notion of collective capabilities and students' possibilities for participating in organized collective action. With regard to methods, future studies are encouraged to employ longitudinal interviews and complement student narratives with staff interviews. Studies exploring a wider range of data collection methods such as focus groups, capability-based student surveys, a review of artefacts, or a combination of these approaches would provide additional layers of analysis. Overall, while a capabilities-based case study is only one of a myriad of ways through which we can examine the links between international education and well-being, this discussion has illustrated the fruitfulness of such an approach when understanding the role of cross-border education in enhancing personal well-being and contributing to a flourishing collective life.

REFERENCES

Anabo, I., I. Elexpuru, and L. Villardón-Gallego. 2019. Revisiting the Belmont Report's ethical principles in internet-mediated research: Perspectives from disciplinary associations in the social sciences. *Ethics and Information Technology* 21, no. 2: 137–149. Doi: 10.1007/s1067 6-018-9495-z.

Bauer, J., D. McAdams, and J. Pals. 2006. Narrative identity and eudaimonic well-being. *Journal of Happiness Studies* 9, no. 1: 81–104. Doi: 10.1007/s10902-006-9021-6.

Boni, A., and M. Walker, eds. 2013. *Human Development and Capabilities. Re-imagining the University of the Twenty-First Century.* London: Routledge.

Bracht, O., C. Engel, K. Janson, et al. 2006. *The Professional Value of ERASMUS Mobility.* Kassel: International Centre for Higher Education Research (INCHER-Kassel).

Brighouse, H., and E. Unterhalter. 2010. Education for primary goods or for capabilities? In *Measuring Justice: Primary Goods and Capabilities*, eds. H. Brighouse, and I. Robeyns, 193–214. Cambridge: Cambridge University Press.

Cai, W., and G. Sankaran. 2015. Promoting critical thinking through an interdisciplinary study abroad program. *Journal of International Students* 5, no. 1: 38–49. https://www.ojed.org/index.php/jis/article/view/441.

Cicchelli, V. 2013. The cosmopolitan 'Bildung' of Erasmus students' going abroad. In *Critical Perspectives on International Education*, eds. Y. Hebert, and A. A. Abdi, 205–208. Rotterdam: Sense Publishers.

Danermark, B., M. Ekstrom, L. Jakobsen, and J. C. Karlsson. 2002. *Explaining Society: Critical Realism in the Social Sciences.* London: Routledge.

Deneulin, S., and J. A. McGregor. 2010. The capability approach and the politics of a social conception of wellbeing. *European Journal of Social Theory* 13, no. 1: 501–519. Doi: 10.1177/1368431010382762.

Easterbrook, M. J., T. Kuppens, and A. Manstead. 2016. The education effect: Higher educational qualifications are robustly associated with beneficial personal and sociopolitical outcomes. *Social Indicators Research* 126, no. 3: 1261–1298. Doi: 10.1007/s11205-015-0946-1.

Evans, P. 2002. Collective capabilities, culture, and Amartya Sen's Development as Freedom. *Studies in Comparative International Development* 37, no. 2: 54–60.

Eurofound. 2017. *European Quality of Life Survey 2016: Quality of Life, Quality of Public Services, and Quality of Society.* Luxembourg: Publications Office of the European Union.

European Commission. 2019. *Erasmus plus programme guide.* https://ec.europa.eu/programmes/erasmus-plus/resources/documents/erasmus-programme-guide-2020_en.

European Commission. n.d. *Scholarship statistics.* https://eacea.ec.europa.eu/erasmus-plus/library/scholarship-statistics_en (accessed January 3, 2020).

Eurostat. 2015. *Quality of Life: Facts and Views.* Luxembourg: Publications Office of the European Union.

Flores-Crespo, P. 2007. Situation education in the human capabilities approach. In *Amartya Sen's Capability Approach and Social Justice in Education*, eds. M. Walker, and E. Unterhalter, 45–65. New York: Palgrave Macmillan.

Hemmer, S., S. Pommer, J. Knabl et al. 2011. Clustering Erasmus Mundus masters courses and attractiveness projects. Lot 2: Employability. Survey results October 2011. http://eacea.ec.europa.eu/erasmus_mundus/clusters/documents/publication_version_employability_survey_results.pdf.

Hsieh, H., and S. Shannon. 2005. Three approaches to qualitative content analysis. *Qualitative Health Research,* 15, no. 9: 1277–1288.

ICU.net AG. 2014. Erasmus Mundus graduate impact survey – September 2014. https://www.em-a.eu/fileadmin/content/GIS/Graduate_Impact_Survey_2014.pdf.

Jongbloed, J. 2018. Higher education for happiness? Investigating the impact of education on the hedonic and eudaimonic well-being of Europeans. *European Educational Research Journal* 17, no. 5: 733–754. Doi: 10.1177/1474904118770818.

Kruger, T., K. Klein, S. Pinkas, A. Hopfner, and J. Kuske. 2017. Erasmus Mundus graduate impact survey 2017. https://www.em-a.eu/fileadmin/content/GIS/GraduateImpactSurvey_2017_final_web.pdf.

Mazzoni, D., C. Albanesi, P. D. Ferreira, et al. 2018. Cross-border mobility, European identity and participation among European adolescents and young adults. *European Journal of Developmental Psychology* 15, no. 3: 324–339. Doi: 10.1080/17405629.2017.1378089.

Mitchell, K. 2012. Student mobility and European identity: Erasmus study as a civic experience? *Journal of Contemporary European Research* 8: 490–518.

Mitchell, K. 2015. Rethinking the 'Erasmus effect' on European identity. *Journal of Common Market Studies* 53: 330–348. Doi: 10.1111/jcms.12152.

Narayan, D., and P. Petesch. 2002. *Voices of the Poor from Many Lands.* Washington, DC: The World Bank.

Nussbaum, M. 2000. *Women and Human Development: The Capabilities Approach.* Cambridge: Cambridge University Press.

Nussbaum, M. 2011. *Creating Capabilities.* Cambridge, MA: The Belknap Press of Harvard University Press.

OECD. 2019. *Education at a Glance 2019: OECD Indicators.* Paris: OECD Publishing. Doi: 10.1787/f8d7880d-en.

Robeyns, I. 2003. Sen's capability approach and gender inequality: Selecting relevant capabilities. *Feminist Economics* 9, no. 2–3: 61–92. Doi: 10.1080/1354570022000078024.

Robeyns, I. 2017. *Wellbeing, Freedom and Social Justice: The Capability Approach Re-examined.* Cambridge: Open Book Publishers.

Saito, M. 2003. Amartya Sen's capability approach to education: A critical exploration. *Journal of Philosophy of Education* 37, no. 1: 17–33.

Saldaña, J. 2013. *The Coding Manual for Qualitative Researchers.* Thousand Oaks, CA: SAGE Publications Inc.

Saumure, K., and L. Given. 2008. Nonprobability sampling. In *The SAGE Encyclopedia of Qualitative Research Methods*, ed. L. Given, 562. Los Angeles, CA: SAGE Publications, Inc.

Sen, A. 1992. *Inequality Reexamined.* New York: Russell Sage Foundation.

Sen, A. 1999. *Development as Freedom.* New York: Alfred A. Knopf, Inc.

Tarrant, M. A., D. L. Rubin, and L. Stoner. 2014. The added value of study abroad: Fostering a global citizenry. *Journal of Studies in International Education* 18, no. 2: 141–161. Doi: 10.1177/1028315313497589.

Terzieva, B., and M. Unger. 2019. *Erasmus Mundus Joint Master Graduate Impact Survey 2018.* Institute for Advanced Studies. https://em-a.eu/fileadmin/content/GIS/GIS2018_report_.pdf.

Walker, M. 2006. *Higher Education Pedagogies: A Capabilities Approach.* Berkshire: Open University Press.

Walker M. 2007. Widening participation in higher education: Lifelong learning as capability. In *Philosophical Perspectives on Lifelong Learning,* ed. D. N. Aspin, 131–147. Dordrecht: Springer.

Walker, M. 2008. Human capability, mild perfectionism and thickened educational praxis. *Pedagogy, Culture and Society* 16, no. 2: 149-162. Doi: 10.1080/14681360802142112.

Wilson-Strydom, M., and M. Walker. (2015). A capabilities-friendly conceptualisation of flourishing in and through education. *Journal of Moral Education* 44, no. 3: 310–324. https://doi.org/10.1080/03057240.2015.1043878

UNESCO. 2016. *Global Education Monitoring Report 2016. Education for People and Planet: Creating Sustainable Futures for All.* Paris: UNESCO.

Yin, R. K. 2018. *Case Study Research and Applications: Design and Methods* (6th ed.). Los Angeles, CA: SAGE Publications, Inc.

Young, M. 2009. Basic capabilities, basic learning outcomes and thresholds of learning. *Journal of Human Development and Capabilities* 10, no. 2: 259–277. Doi: 10.1080/19452820902941206.

5 Improving Quality of Life through Global Citizenship Education (GCED)

Ashita Raveendran
National Council of Educational
Research & Training (NCERT)

CONTENTS

5.1 INTRODUCTION

Education with its transformative power helps in evolving the learner encompass the skills and values which are necessary for building in a quality life. It is not just about literacy and numeracy, the learners also acquire the capability to fight against injustice, violence, corruption etc. in quest of unity and with respect to diversity. Education shapes the individual learner with the Socio-Emotional Competencies (SECs) required for creating inclusive, peaceful, just and sustainable societies (UNESCO 2014a).

In the present situation of global challenges and issues like threat of war, terrorism, human rights violations, climate change, refugee crisis, inequalities, poverty etc., attainment of quality of life remains to be a distant reality. The global peace, harmony and sustainability are hardly hit by these challenges affecting all nations irrespective of their development. It is in response to these challenges and realizing the role education can play in alleviating the quality of life, the UNESCO proposed for Global Citizenship Education (GCED) which is also reflected as Target 4.7 in the Sustainable Development Goals (SDGs) of United Nations General Assembly (UNESCO 2016). The intention is to make the learners understand the global issues and enable them to become active members of society, connect with nature and promote peaceful co-existence that in turn will help in transforming the world into more peaceful, tolerant, inclusive, secure and sustainable place to live in.

The paper throws light on how quality of life can be impacted and influenced by education. It stresses on the need for imparting education that can promote human values and make the learners an individual socio-emotionally competent. Bringing in the importance of GCED, the paper analyses how far the curricular practices related to GCED have influenced the SEC of school students. Being socio-emotionally competent will help in generating awareness on law and order, and address issues like social crimes, communal harmony and substance abuse. This will help in making them ready to contribute to the society, by allowing them to apply many important values in life.

5.2 QUALITY OF LIFE AND EDUCATION

Quality of Life as put forth by WHO is an individual's discernment of their position in life in the milieu of the culture and value systems in which they live relative to their goals, expectations, standards and concerns. The domains of quality of life include physical health, psychological state, level of independence, social relations, environment, spirituality and personal beliefs (WHO 1997) which exists jointly with the others. Education paves the way for improvement in the quality of life by providing access to the primary determinants of well-being: employment, socio-economic status and psychological and physical health. Employment apart from the monetary benefits also offers people the freedom from boredom, and external control and also an opportunity to use their skills and competencies, thereby increasing subjective well-being.

Education can also pave the way for possession of positive values, which in turn boosts apparent control on emotional and physical distress, thereby improving the subjective quality of life, measured as psychological well-being and distress. Education makes a person wise enough to make use of the knowledge attained for good cause and live judiciously with the environment. Thus, education transforms lives, making it the main driver of development, and also supports for the attainment of the SDGs. The SDG 4 –Target 4.7 calls for all learners to acquire the knowledge, skills and values needed to promote sustainable development, a blue print for a better world. The New Education Policy (NEP) 2020, in India supporting the SDG 4.7, also has called for concerted curricular and pedagogical initiatives in GCED (Government of India 2020).

Looking at education in the context of quality of life, it is seen to have influenced the objective and the subjective quality of life. Being the main driver of technological innovation and high productivity, educational attainment improves the likelihood of finding a job, earning a standard income, increased lifetime earnings, all leading towards a good quality of life. The limited access to skills and competencies limit access to labour market putting the workforce at risk. It often makes them prone to unemployment, increases the risk of social exclusion and poverty and also may hinder their participation in social and political activities.

Not denying the need for sharpening intelligence and acquiring corpus of knowledge for success in life, inculcating values and social–emotional skills that are fundamental to corporeal and mental well-being, remains to be a requisite for accomplishing a meaningful life. For any country, it is not the growth in gross domestic product that seems to be most important determinant in deciding the quality of life. Even if all its citizens are rich, if they are not good-hearted human beings, lack social skills and are unable to control emotions, all amounting to dissatisfaction and unhappiness, the nation remains to be poor with respect to the quality of life. Education is not just an enduring progression for improving knowledge and skills, but also a unique means of bringing about personal development and edifying relationships amid individuals, groups and nations (UNESCO 1996). It is not theoretical knowledge and artificial intelligence that makes a good-hearted human being, it's the personality development, social skills, cooperative learning practised in schools as a part of the socio-emotional learning that contributes towards it.

5.3 ASSOCIATION OF SOCIO-EMOTIONAL COMPETENCE (SEC) WITH LIFE OUTCOMES

The SECs regulates one's thoughts, emotions and behaviour which can influence many important life outcomes (OECD 2015, 2018) and even influence the decision-making which is a process that emerges from the convergence of rational and emotional brain (Lerner et al. 2015) The emotional behaviour expressed are not just part of the natural reactions. They can be changed/controlled through adequate training. The SECs include the "ability to recognize and appreciate different and multiple identities and common humanity; develop skills for living in an increasingly diverse world; develop attitudes of care and empathy for others and the environment; respect for diversity; develop values of fairness and social justice and skills to critically analyze inequalities based on gender, socio-economic status, culture, religion, age, and other issues" (APCEIU 2018). The higher trait of emotional competence provides greater well-being and higher self-esteem (Schutte, Malouff, Simunck, McKenley, & Hollander 2002). Emotional competency makes individuals capable of emotional regulation and is less prone to psychological disorders (Gross & Munoz 1995) and rarely adopts unhealthy behaviours (Brackett, Mayer, & Warner 2004). The acquiring of SECs will help in enabling them to calm down or overcome their negative emotions like fear, anger, sadness, rejection and convert it into positive thoughts and emotions. Being socio-emotionally competent helps in generating awareness on law and order, resolving conflicts respectfully, decision-making, making ethical and safer choices and also addressing issues like social crimes, communal harmony and substance abuse. This will help in making the learners ready to contribute to the society by allowing them to apply many important values in life.

5.3.1 SEC FOR ACADEMIC SUCCESS

Studies have shown that the SECs acquired by the leaner is also linked with the better academic performance (Brackett, Elbertson, & Rivers 2016; Weissberg, Durlak, Domitrovich, & Gullotta 2015; Zins, Bloodworth, Weissberg, & Walberg 2004), and conversely, lack of SECs can result in poor academic performance (Hawkins, Farrington, & Catalano 1998). The programmes related to socio-emotional learning are seen to have addressing positive development including improving academic performance and reducing the risk of academic failure (Catalano, Berglund, Ryan, Lonczak, & Hawkins 2002). In addition to fewer behaviour problems and less emotional distress, the learners with SEC demonstrated improved attitude towards academics and significant improvement in academic achievement (Durlak, Weissberg, Dymnicki, Taylor, & Schellinger 2011). The emotional distress can interfere with the cognitive processes hampering attention, remembering key concepts and completing academic work (Merrell & Gueldner 2010) which may also affect the development of intrinsic enthusiasm essential for long-term engagement in studying and intellectual pursuits (Zins, Bloodworth, Weissberg, & Walberg 2004). Conversely, learners showing better academic performance are also seen to possess positive socio-emotional behaviour (Payton et al. 2008). Even when the academic and socio-emotional skills are seen to be reinforcing one another, the education system hesitates to undertake such learning programmes as it does not envisage clear and obvious benefits which is reflected in the form of grades/marks.

5.3.2 SEC FOR CAREER BUILDING

Beyond the cognitive abilities of learning to know and learning to do, the contemporary world demands for learning "how to be" and learning "how to behave". Learning on how we perceive and live with others in a social context facilitate and promote employability (Hettich 2000; Gharoie 2011). One's attitude towards work, towards other fellow beings, ability to adapt to situations, being flexible all hold relevance in career contexts. Professional success necessitates skill to work in team, enter into negotiations, be work stress tolerant, be capable of conflict solving, address ethical issues and handle critical and adverse situations. Evidences show relationship between SECs and career decision-making (Di Fabio, 2011; Caruso & Wolfe 2001). The competence to manage work stress and pressure affects the work performance (Bachman et al. 2000) and is also a requisite for effective leadership (Wolff, Pescosolido, & Druskat 2002). The socio-emotionally competent individual will be able to withstand the adversities in life and work that range from unemployment, stagnation in profession, low wages, hierarchy, complex socio-labour relationships etc. Emotional competence stands to be a key element in finding a job. Including emotional competence modules in the education can increase employability and also impact personality development (Nelis et al. 2011). The attainment of better career/job results in improved socio-economic status which contributes to psychological well-being. The SECs acquired by the individual during the learning process are capable of addressing the adverse effects of instability in jobs, insecure socio-labour relationships and increasing quality of life at work.

5.3.3 SEC FOR PERSONAL WELL-BEING

While the cognitive skills can bring in academic success through better job opportunities, the SEC remains to be essential for personal accomplishment (Richardson & Evans 1997). Studies show the importance of socio-emotional learning for children's mental health, social relationships and school performance (Greenberg et al. 2003) and that it also prevents some important negative outcomes (Weissberg, Durlak, Domitrovich, & Gullotta 2015) which was further established in the study on adults by Nelis et al. (2011). The scores of the learners with respect to the five domains of SEC is positively related to their academic performance and the strengthening of such curricular practices positively influences the capacity to learn and also enable the learners to experience personal satisfaction (Ee, Cheng, & Zhou 2012). Researches in psychology and on neuroscience show the origins of violence, aggression and hatred to be driven by perceptions of threat, estrangement and negative emotions (McDermott & Hatemi 2013), which can be trained using behavioural tools of socio-emotional learning.

5.3.4 SEC FOR GLOBAL WELL-BEING

The current social and political environment calls for urgency of global awareness and learning to live together in a challenging world. In order to thrive in an interconnected, diverse and globalized world, learners need to be empowered with the SECs. Acquiring of the SECs bring in improved performance in school and ultimately in life (CASEL 2013). The GCED and Education for Sustainable Development (ESD) can help in eradicating extremism and terrorism, the biggest challenges in today's world. It can foster the principles of universal living and promote the values of acceptance and learning to live together.

5.4 GLOBAL CITIZENSHIP EDUCATION (GCED) AND THE SOCIO-EMOTIONAL COMPETENCE (SEC)

Education is not confined to just reading, writing or learning of concepts. It needs to ensure development of skills, values and attitudes, learn to resolve problems assertively, to be empathetic and sensitive. There should be space for enhancing the cognitive, socio-emotional and behavioural aspects of the learner. All the four pillars of learning, namely (1) learning to know, (2) learning to do, (3) learning to be and (4) learning to live together (Delors et al. 1996), are relevant to bring in objective and subjective changes in the quality of life. While the pillars learning to know and the learning to do will ensure objective quality of life by way of providing work, income and a better standard of living, the pillars learning to be and learning to live together is crucial in enabling individual and societies to live in peace and harmony. In this age of globalization, possession of social skills and values are required for a better quality of life.

5.4.1 GCED AND ITS IMPLICATIONS

The GCED aims to empower learners to assume active roles to face and resolve global challenges and to become proactive contributors to a more peaceful, tolerant, inclusive and secure world (APCEIU 2018). The GCED imparted through various curricular practices helps learners appreciate difference and multiple identities; gather understanding of global governance structures; make connections between global, national and local systems and processes; develop skills for living in an increasingly diverse world; develop and apply critical thinking skills for civic literacy; develop attitudes of care and empathy for others and the environment; respect for diversity; develop values of fairness and social justice; critically analyse inequalities existing in the society; and address global issues as informed, engaged, responsible and responsive global citizens (UNESCO 2015).

With the changes occurring in the economic, cultural, technological fields, the world is getting more connected and people more interdependent. These are making the challenges, climate change, conflicts and war, gender discrimination, environmental degradation, sustainable development, racism, terrorism, etc. irrespective of its place of existence, global in nature, affecting all. For minimizing the damages of global issues, facing global challenges and optimizing the benefits, we need to make our citizens capable of participating in local national and global civic life. GCED will help in developing a sense of global citizenship among the students. The Global Education First Initiative (GEFI) in 2012 had made fostering global citizenship as one of its three education priorities and since then UNESCO has been promoting GCED.

The UNESCO 2015, Guidance on GCED sets out three core dimensions or learning domains:

  i. **Cognitive**: to acquire knowledge, understanding and critical thinking about global issues and the interconnectedness/interdependency of countries and different populations;
 ii. **Socio-emotional**: to have a sense of belonging to a common humanity, sharing values and responsibilities, demonstrating empathy, solidarity and respect for differences and diversity; and
iii. **Behavioural**: to act responsibly at local, national and global levels for a more peaceful and sustainable world.

In India, there exist various curricular practices which directly or indirectly address these global issues and also help in developing the SECs. The curricular practices that focus on the GCED can enable learners to possess the SECs along with the cognitive skills which will develop them as well-informed, responsible, helpful, industrious, peaceful, ethical and duty abiding citizens (Elias et al. 1997). GCED "highlights essential functions of education related to the formation of citizenship [in relation] with globalization". (UNESCO 2014b). GCED becomes one of the important components of school education all over the world that will help to not only teach people to aspire for a "good life" as an individual but to hope goodness and righteousness for all of humankind and the world.

5.4.2 Interventions in Curricula & the Change Brought in by GCED

In India, the NEP 2020 is the first policy document in the school curriculum that has explicitly talked about GCED. However, there are various curricular practices that are being implemented in India, that foster value inculcation and citizenship education and that consist of the elements of GCED. An analysis of the curricular content shows evidence of the importance given for imbibing values and enhancing the SEC in the learners. There are various programmes that are being carried out by different NGOs, state governments, civil society, religious cult organization, etc. which promotes GCED. For instance, the *Vidyavahini* project was implemented in schools in the state of Sikkim, with the help of school administration and Sri Sathya Sai Sewa organization. Under the *Vidyavahini* project, the teachers are enabled to conduct activities that help to inculcate moral and social values among the students. Similarly, the Student Police Cadet (SPC) project undertaken in the state of Kerala is also one of the best examples which implicitly helps in inculcating the elements of GCED. The project focuses on inculcating discipline, decision-making and integrity of the students. Other programmes like "*Mulyavardhan*" in Maharashtra, "*Saptadhara*" in Assam and Gujarat, "Happiness Curriculum" in Delhi, "*Pratibha Parv*" in Madhya Pradesh, "Joyful Saturday" initiative in Rajasthan, etc. are some of the programmes that are implemented as a part of the school curriculum all of which enhances the skills and competencies of the learner and are very much related to GCED.

The research findings of the study undertaken by NCERT on the various curricular practices show that this kind of practices ensure active participation of students in projects that address the social, political, economic, or environmental issues stretching from local to the global (Subhash and Ashita 2019). The curricular practices have substantively changed the attitude and behaviour of most of the students and make them responsible members of society. The research study was undertaken during the period April 2018–March 2019, in order to find out the relation between curricular practices related to GCED and SEC among the secondary school students.

Findings of the analysis with respect to the SPC Project as a curricular practice having the elements of GCED have been put forth in Section 1.4.3 to show how the GCED practices can help in influencing the SEC.

5.4.2.1 Methodology

Research Question
- To what extent the curricular practices followed in the schools are capable of enhancing SEC among students to act as global citizens?

Hypothesis
Ha: The SPC students would yield significant score across the SECs relative to the comparison group the non-SPC students.
Sample of the study and tools: The SPC Project was being implemented in selected schools of Kerala and was offered to students who were willing. The SPC and the non-SPC groups were composed of 95 students each in both the groups belonging to three schools in the state of Kerala where the

SPC programme have been implemented. There was no evident difference
in the sociodemographic characteristics between the students and all the
students selected were studying in the class IX (Age group 13–15).

The socio-emotional scale developed contained a pool of 40 items with eight
items for each component classified under the five domains: self-awareness;
self-management; social awareness; relationship skills and responsible decision-
making. Each item selected for the competency scale reflected one of the five
domains. The scale items include: "I play nicely with others" (self-awareness); "I
have control over my emotions" (self-management); "It is easy for me to understand
why people feel the way they do (social awareness); "I would do my best to be there
for a friend who need me" (relationship skills); and "When making decisions, I
take into account the consequences of my actions" (responsible decision-making).
Students rated each item on a 3-point scale (1 = *not at all true of me*, 2 = *somewhat
true of me*, 3 = *very true of me*). Piloting of the tools was conducted for verifying
the reliability and validity of the measure. The SEC was assessed using the self-
report items from the learners which was triangulated with the assessment made by
teachers, parents and peer group.

5.4.3   THE STUDENT POLICE CADET (SPC) PROJECT – A CURRICULAR INTERVENTION

The SPC Project, a school-based youth development initiative, initially implemented
in the state of Kerala jointly by the Departments of Home and Education and supported by Departments of Transport, Forest, Excise and Local Self-Government, has
been launched nationally and is been implemented in many states in India. It trains
the learners

> to evolve as future leaders of a democratic society by inculcating within them respect
> for the law, discipline, civic sense, empathy for vulnerable sections of society and resis-
> tance to social evils. The project also enables youth to explore and develop their innate
> capabilities, thereby empowering them to resist the growth of negative tendencies
> such as social intolerance, substance abuse, deviant behavior, and anti-establishment
> violence

> *(Student Police Directorate 2017)*

which are all aligned with GCED practices.

The SPC project provides the field-level experiences that can facilitate positive
change which can be valuable to society in the long run and create substantial positive takings. The practices seem to be capable of enhancing SECs among learners
to act as global citizens. The SEC scale of the students who had undergone the SPC
programme in comparison with the non-SPC students reveal that the learners who
were involved in the programme demonstrated confidence and discipline.

The studies on SECs have analysed the ability of an individual to control one's
emotions (Eisenberg et al. 1997); manage peer relations (Rubin & Parker 2006);
and promote resilience (Cowen, Wyman, & Work 1996; Eisenberg et al. 1997;

Elias et al. 1997). Five intrapersonal and interpersonal levels of skills, viz., self-awareness, self-management, responsible decision-making, relationship management and social awareness have been identified (Zins, Bloodworth, Weissberg, & Walberg 2004; CASEL 2017) to evaluate the SECs. The assessment in the schools need to focus on using appropriate tools for ascertaining the SECs along with the academic achievement.

Figure 5.1 shows the results of the SEC score of the SPC and non-SPC students prepared on the basis of data collected from the schools of Kerala state, India. The analysis of the SEC scale shows that there is a high score for decision-making skills, followed by self-awareness, social awareness with the self-management scoring the least for the SPC students. The average score of SEC for an SPC student stands at 96.46, whereas for a non-SPC student, it is 88.87. The average SEC score of SPC student in comparison with the non-SPC student shows the effect of the SPC project on increasing the competencies. The results show a significant greater score in the emotional domain of the SPC group in comparison with the non-SPC. More specifically, findings indicated that the SPC project was able to produce significant positive effects on the five domains of SECs.

Being self-aware makes the learner to be reflective and clear about the reasons for their emotional responses and be able to regulate their behaviour (Froming and McManus 1998). It makes them possible to possess self-control of their emotions and take more responsible decision-making in life. The SEC score of non-SPC students show that they possess low relationship skills.

Self-management is requisite for being successful in peer relationships, sustaining close relationships, succeeding at work and maintaining physical health (Hubbard & Coie 1994). Scores with respect to the competencies self-management and social awareness remain to be the same for the non-SPC students and much lower when compared to the SPC students. Differences, even though minor, are also seen in the self-awareness and self-management competencies between the SPC and the non-SPC students. The self-management and self-regulation help the learners in controlling

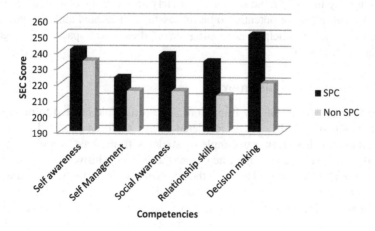

FIGURE 5.1 Socio-emotional competency score of SPC and non-SPC students.

emotional outbursts and impulsive reactions, be engaged in the class, interact in appropriate manner with teachers and others (Zins, Bloodworth, Weissberg, & Walberg
2004) and also avoid substance use and violence (Payton et al. 2008).

Responsible decision-making ability is a prerequisite for a rational human being
and for contributing to the well-being of community. The SPC project undertakes
various activities which are mostly done in groups. Teamwork helps in developing the competency required for decision-making. Relationship management with
the peers is important not just while engaging at school, but also in social living.
Empathy has close linkages with social living and awareness. Being empathetic
makes an individual capable to read other persons' cues and to understand, and suit-
ably respond to their emotional state (Frey & Guzzo 2000). Loneliness due to lack
of relationship skills can make individuals detached from social living and make
oneself depressed adversely affecting their psychological state of mind. The scores of
the SECs show that the activities undertaken under the curricular practice – SPC programme, have contributed towards attainment of the social awareness, relationship
skills and decision-making skills within the learners.

Consistent with the hypothesis, the results indicated that the students who had
undergone the SPC programme demonstrated better SECs compared to their counterparts who were not SPCs. The SPC students demonstrated enhanced SECs. This
programme which had elements of GCED have exerted effects on the self-awareness,
self-management, social awareness, relationship skills and decision-making among
the students. These findings are also consistent with the other studies which conclude
that GCED programmes influence the SECs.

The aim of SPC project being training students to evolve as future leaders of a
democratic society by inculcating in them respect for law, discipline, civic sense,
empathy for vulnerable sections of society and resistance to social evils has been
successful to a larger extent, as evidenced by its high score of SECs. The teachers
also appreciated the SPC project stating its positive effects in the academic performance, school atmosphere and the changes it brings in their students' life. The
students were nurtured with the qualities required for good citizens, with better civic
sense, empathy for vulnerable sections of society and resistance to social evils. The
SPC programme has the potential to be successfully integrated and implemented in
the schools which will help in generating nurtured citizens impacting their general
and total quality of life.

5.4.4 Limitations of the Study

The study is not devoid of any limitations. The SEC scale created to measure the
score is based on CASEL and is restricted to five socio-emotional domains. The
study results are based on the sample of students from three schools which had
undertaken the SPC programme, hence maybe less conclusive. As the study was
not a longitudinal one, the changes in the competencies over the years could not be
ascertained. Eliciting valid responses from a diverse population within the limited
time period of interaction was difficult. Even though the study results were correlated
with the observations and interviews made by the researcher, there was lesser scope
for interacting with parents and the community.

5.5 SCOPE FOR FURTHER RESEARCH & IMPLICATIONS FOR EDUCATION

These limitations lead us to suggest a roadmap for future research on socio-emotional skills in the educational context of GCED. Numerous questions and gaps remaining, it remains binding upon researchers to trace out the data to fill in the gaps and bring the indistinct facets into focus. In this globalized world full of competition, growing uncertainty and mounting risks, in addition to the rise in intolerance, violence and extremism (Institute for Economics & Peace. Global Peace Index 2020), there are reports and cases of increased anxiety, stress and depression among the people all over the world (World Health Organization 2017). The education systems are not necessarily addressing the socio-emotional well-being of its learners which is a cause of concern. The inclusion of age-appropriate programmes that address social and emotional skills such as empathy and self-regulation are important milestones (Denham et al. 2016) that are connected with a range of important later-life outcomes, including mental health and well-being, health and health behaviours, family and relationship stability and labour market success (Goodman et al. 2015). Such programmes are evidenced to promote well-being, social–emotional skills, along with cognitive skills, thereby preventing mental health lifetime risks and enhance learner's success both in school and life (Diamond & Lee 2011; Durlak Roger, & Dymnicki 2011; Greenberg & Harris 2012).

The increasingly strong focus laid on academic learning largely decided by the test scores has limited the scope of education. Even with the policies and curriculum frameworks advocating for the holistic and socio-emotional development of students (NCERT 2005), the schools, teachers, parents all are engaged in the mad race of preparing the learners for scoring high marks. Attainment of social-emotional competence that helps the learner's adjustment into adulthood (Merrell & Gueldner 2010) are not often adequately addressed within the school curriculum.

The global challenges of poverty, inequality, food insecurity, war and terrorism will continue on its rise and remain unresolved until accelerated progress towards quality education remain unattained. The curriculum needs to tap the talents and potential within the learners, bring in changes in their attitudes and personalities so as to enable them to not only to improve their lives but also transform societies. Taking into consideration all these whys and wherefores, it is pertinent to design and implement interventions that can develop such competencies into the school curriculum and in teaching-learning strategies, which shall remain with the learner throughout life. The policies and practices need to place value on respecting human dignity and human rights, focus on cultural diversity, be inclusive, integrate knowledge of other people, places and perspectives, be adaptive to meet the needs of diverse learners be capable of dealing with multi-grade classrooms, develop appropriate curricular materials, help in empowering educators and find place for dealing with contentious topics that can reinforce empathy and respect (UNESCO 2014a)

These measures implemented holistically will not only augment the learner's emotional vocabulary, but also help them to acquire strategies to manage emotionally trying circumstances, accomplish self-control, enhance the tolerance level and control conflictive desires. At the global level, interest in developing competencies in academic settings has risen, as is seen in the GCED being advocated as a part of the SDG4.7 target, programmes and advocacy material brought out by UNESCO, APCEIU, MGIEP, CASEL etc., whose aim is to develop generic and specific competencies among the learners (MGIEP: UNESCO 2017). The Global Education Monitoring (GEM) Report of 2016 which reviewed over 110 national curriculum framework documents found that among the 78 countries, 92% of them included human rights in their curricula, about 73% and 62% of the countries were found to have included sustainable development and peace, respectively (UNESCO 2016).

Various forms of transformative education, viz., Disarmament Education, Peace Education, Education for a Culture of Peace, Human Rights Education, Education for International Understanding, Education for Global Justice, Civic education, Education for Democratic Citizenship, Values Education, Life Skills Education, ESD, GCED, etc. have been advocated at different time periods and are in existence in the global arena at various stages, all of which share common visions, goals, principles and values (APCEIU 2017).

Interventions that can help to inculcate SECs that will exist throughout life, need to be integrated into the school curriculum. This not only will augment their emotional competency, but also help to acquire skills to cope in emotionally difficult situations, attain emotional restraint, so as to effectively manage emotions and conflictive desires (Fredrickson 2001). Analysis of textbooks that reflect the classroom reality show an up rise in inclusion of human rights, gender equality, environment, etc. However, this mainstreaming of ESD and global citizenship does not seem to be sufficient. It is very evident that all education cannot bring in the required changes. We need to bring in the scope for implementing transformative education that can develop a culture of global peace and well-being.

The curriculum, to create learners instilled with the ethos, needs to be such that (1) it gives importance to respecting human dignity and valuing human rights; (2) respecting varied diversity by emphasizing it as a human right; (3) be inclusive in nature; (4) integrates knowledge of other people and perspectives; (5) provide competencies to protect themselves as well as rights of others; (6) provides positive cases of diverse ethnic groups and (7) is capable of dealing with controversial issues without conflict (UNESCO 2014a). Promoting GCED can help fostering the competencies and skills required to prepare young learners to be globally competent. The acquiring of global competence is necessary for employability, for living cooperatively in diverse situation, to communicate and behave responsibly and eventually for attainment of the SDGs. The curricular materials, teaching-learning, assessment and all the activities within and outside the classroom needs to focus on transforming the young learner into a global citizen possessing integrity, ethical and respect for rule of law that can benefit in maintaining global peace and prosperity.

5.6   CONCLUSIONS

For the dream of global peace to come true, in a world perpetually on the edge of war, our curriculum needs to focus on the promotion of global citizenship. We live in a dire state where the natural resources are exploited, climatic changes, all air, water and land remain polluted, poverty with widening inequality, added with natural calamities, food insecurity, displaced and distraught families, etc. all of which are adversely affecting the quality of life. These concerns of the day point on to the need for sustainable development for safeguarding human well-being. Well-being of an individual as we see, is impacted by education through improved job opportunities, economic conditions, supportive social structure and family conditions.

Likelihood of attaining quality jobs, income etc. increases with the attainment of education, resulting in the uplift in social status. However, just being educated and in job may not help in achieving a good quality of life. It may objectively fetch one with a paid work but more important is the attainment of the subjective goals. It is only when the benefits of growth reach all the citizens and the environment, there will be improvement in the quality of life. Technological innovations, higher income growth and better standard of living when limited to a small percentage of the population may not necessarily contribute towards the quality of life. One cannot live happily for long in the world full of sufferings, poverty and misery. Educational systems preparing the learners for just work/jobs, that will help in meeting the economic ends only, ends in making of an unsustainable dead world.

Education needs to develop citizenship education competencies, including values and diversity in society, promoting a sense of shared responsibility, aspiring for goodness and justice for the whole world, thereby enhancing the quality of life and sustainability. The curricular practices have an extensive role to play in making the students acquire SECs enabling them to act as global citizens. The hypothesis of acquiring significant score through the curricular practices that focused on the SECs has been validated in the study. The education sector needs to mould itself to reap the fruits of globalization and capture the opportunities available with its advent. Policies need to be devised to enhance the SECs, which in turn, improves one's quality of life both objectively and subjectively. The GCED, with its potential in increasing the SECs and attaining socio-psychological resources of sense of control and social support, offers a promising approach and can go a long way in the attainment of Quality of Life.

ACKNOWLEDGEMENTS

The research conducted as a part of a larger study titled "A study on Global Citizenship Education in India", undertaken with the PAC funding of the National Council of Educational Research and Training (NCERT), an autonomous institution under the Ministry of Education, Govt. of India, have helped in determining the influence of GCED practices. The author also acknowledges the deeper insights gained from the fourth Global Capacity Building Workshop on the GCED organized by the Asia-Pacific Centre of Education for International Understanding (APCEIU) Republic of Korea partnered with UNESCO. The author is thankful to the authorities concerned for providing the opportunities and also the teachers, students and other stakeholders with whom the author had an opportunity to interact. A thank note is also due to the reviewers for their incisive comments.

REFERENCES

APCEIU. 2017. *Global Citizenship Education- A Guide for Policy Makers.* Seoul: APCEIU.
APCEIU. 2018. *Global Citizenship Education A Guide for Trainers. Office of Education and Training.* Seoul: APCEIU.
Bachman, J., Stein, S., Campbell, K., & Sitarenios, G. 2000. Emotional intelligence in the collection of debt. *International Journal of Selection and Assessment*, 8 (3): 176–182. Doi: 10.1111/1468-2389.00145 (accessed July 22, 2020).
Brackett, M. A., Mayer, J. D., & Warner, R. M. 2004. Emotional intelligence and its relationship to everyday behavior. *Personality and Individual Differences*, 36: 1387–1402. Doi: 10.1016/S0191-8869(03)00236-8 (accessed July 20, 2020).
Brackett, M. A., Elbertson, N., & Rivers, S. 2016. Applying theory, the development of approaches to SEL. In *Handbook of Social and Emotional Learning*, eds. J. A. Durlak, C. E. Domitrovich, R. P. Weissberg, & T. P. Gullota, 20–32. New York, NY: Guilford Publications.
Caruso, D. R., & Wolfe, C. J. 2001. Emotional intelligence in the workplace. In *Emotional Intelligence in Everyday Life: A Scientific Inquiry*, eds. J. Ciarrochi, J. P. Forgas, & J. D. Mayer, 150–167. New York, NY: Psychology Press.
CASEL (Collaboration for Academic, Social and Emotional Learning). 2013. *Effective Social and Emotional Learning Programs- Preschool and Elementary School Edition.* http://casel.org/guide (accessed July 22, 2020).
Catalano R. F., Berglund, M. L., Ryan, J. A., Lonczak, H. S., & Hawkins, J. D. 2004. Positive youth development in the United States: Research findings on evaluations of positive youth development programs. *The Annals of the American Academy of Political and Social Science*, 591: 98–124 Sage Publications. https://jjrec.files.wordpress.com/2016/04/catalano2004.pdf (accessed July10, 2020).
Cowen, E. L., Wyman, P. A., & William, W. C. 1996, February. Resilience in highly stressed urban children. Concepts and findings. *Bulletin of the New York Academy of Medicine*, 73: 267–284.
Delors, J., In'am Al M., et al. 1996. Learning: The treasure within. *Report to UNESCO of the International Commission on Education for the Twenty-First Century.* Paris: UNESCO.
Denham, S. A., Ferrier, D. E., Howarth, G. Z., Herndon, K. J., & Bassett H. H. 2016. Key considerations in assessing young children's emotional competence. *Cambridge Journal of Education.* Doi: 10.1080/0305764X.2016.1146659 (accessed July 22, 2020).
Diamond, A., & Lee, K. 2011. Interventions and programs demonstrated to aid executive function development in children 4-12 years of age. *Science*, 333: 959. Doi: 10.1126/science.1204529 (accessed July 22, 2020).
Di Fabio, A. 2011. Emotional intelligence: A new variable in career decision-making. In *Emotional Intelligence - New Perspectives and Applications.* ed. A. Di Fabio, 51–66. Croatia: InTech Publishers.
Durlak, A. J., Roger, W., & Dymnicki, A. 2011. The impact of enhancing students' social and emotional learning: A meta-analysis of school-based universal interventions. *Child Develoment*, 82(1): 405–432 Doi: 10.1111/j.1467-8624.2010.01564.x (accessed July 22, 2020).
Durlak, J. A., Weissberg, R. P., Dymnicki, A. B., Taylor, R. D., & Schellinger, K. 2011. The impact of enhancing students' social and emotional learning: A meta-analysis of school-based universal interventions. *Child Development*, 82: 405–432. https://www.casel.org/wp-content/uploads/2016/08/PDF-3-Durlak-Weissberg-Dymnicki-Taylor-_-Schellinger-2011-Meta-analysis.pdf (accessed July 10, 2020).
Ee, J., Cheng, S. K., & Zhou, M. 2012. Which mediates achievement: Social emotional competencies or motivational orientations? *Paper Presented at the AARE - APERA Joint Conference*, Sydney, Australia https://repository.nie.edu.sg/bitstream/10497/15505/1/AARE-APERA-2012-EeJ_a.pdf (accessed August 08, 2020).

Eisenberg, N., Guthrie, I. K., Fabes, R. A., Reiser, M., Murphy, B.C., Holgren, R., Maszk, P., & Losoya, S. 1997. The relations of regulation and emotionality to resiliency and competent social functioning in elementary school children. *Child Development,* 68 (2): 295–311.

Elias, M. J., Zins, J. E., et al. 1997. *Promoting Social and Emotional Learning: Guidelines for Educators.* Alexandria, VA: Association for Supervision and Curriculum Development.

Fredrickson, B. L. 2001. The role of positive emotions in positive psychology-the broaden-and-build theory of positive emotions. *American Psychology,* 56 (3): 218–226.

Frey, K. S., Hirschstein, M. K., & Guzzo, B. A. 2000. Second Step: Preventing aggression by promoting social competence. *Journal of Emotional and Behavioral Disorders,* 8 (2): 102–112. Doi: 10.1177/106342660000800206 (accessed July 22, 2020).

Froming, W. J., Nasby, W., & McManus, J. 1998. Prosocial self-schemas, self-awareness, and children's prosocial behavior. *Journal of Personality and Social Psychology,* 75 (3): 766–777. Doi: 10.1037/0022-3514.75.3.766 (accessed July 20, 2020).

Gharoie, A. R. 2011. Emotional intelligence: The most potent factor of job performance among executives. In *Emotional Intelligence - New Perspectives and Applications,* ed. A. Di Fabio, 121–138. Croatia: InTech Publishers.

Goodman, A., Joshi, H., Nasim, B., & Tyler, C. 2015. *Social and Emotional Skills in Childhood and Their Long-Term Effects on Adult Life.* London: Institute of Education.

Government of India. 2020. *National Education Policy, 2020.* New Delhi: Ministry of Education.

Greenberg, M. T., & Harris, A. R. 2012. Nurturing mindfulness in children and youth: Current state of research. *Child Development Perspectives,* 6: 161–166 Doi: 10.1111/j.1750-8606.2011.00215.x (accessed July 22, 2020).

Greenberg, M. T., Weissberg, R. P., O'Brien, M. U., Zins, J. E., Fredericks, L., Resnik, H., & Elias, M. J. 2003. Enhancing school-based prevention and youth development through coordinated social, emotional, and academic learning. *American Psychologist,* 58(6–7): 466–474. Doi: 10.1037/0003-066X.58.6-7.466 (accessed July 10, 2020).

Gross, J. J., & Munoz, R. F. 1995. Emotion regulation and mental health. *Clinical Psychology Science and Practice,* 2 (June) 151–164. Doi: 10.1111/j.1468-2850.1995.tb00036.x (accessed 20 July 2020).

Hawkins, J. D., Farrington, D. P., & Catalano, R. F. 1998. Reducing violence through the schools. In *Violence in American Schools- a New Perspective,* eds. D. S. Elliott, B. A. Hamburg, & K. R. Williams, 188–216. New York: Cambridge University Press.

Hettich, P. 2000. Transition processes from college to career. *Paper Presented at the Annual Conference of the American Psychological Association,* Washington, DC: APA https://files.eric.ed.gov/fulltext/ED447368.pdf (accessed July 20, 2020).

Hubbard, J. A., & Coie, J. D. 1994. Emotional correlates of social competence in children's peer relationships. *Merrill-Palmer Quarterly,* 40(1): 1–20.

Institute for Economics & Peace. 2020. *Global Peace Index 2020: Measuring Peace in a Complex World.* Sydney. http://visionofhumanity.org/reports (accessed on 5 August 2020).

Lerner, J. S., Li, Y., Valdesolo, P. C., & Kassam, K. S. 2015. Emotion and decision making. *Annual Review of Psychology,* 66(1): 799–823 Doi: 10.1146/annurev-psych-010213-115043 (accessed August 02, 2020).

McDermott, R., & Hatemi, P. K. 2013. Threat perception and deterrence. In *Topics in the Neurobiology of Aggression: Implications to Deterrence.* eds. D. DiEuliis & H. Cabayan. Strategic Multi-layer (SMA) Periodic Publication. Feb. 52–62 http://www.antoniocasella.eu/archipsy/Sapolsky_2013.pdf (accessed July 22, 2020).

Merrell, K. W., & Gueldner, B. A. 2010. *The Guilford Practical Intervention in the School's Series. Social and Emotional Learning in the Classroom: Promoting Mental Health and Academic Success.* The Guilford Press. https://psycnet.apa.org/record/2010-09986-000 (accessed July 20, 2020).

MGIEP: UNESCO. 2017. *Rethinking Schooling for the 21st Century: The State of Education for Peace, Sustainable Development and Global Citizenship in Asia*. New Delhi: UNESCO MGIEP.

NCERT. 2005. *National Curriculum Framework 2005*. New Delhi: National Council for Educational Research and Training.

Nelis, D., Kotsou, I., Quoidbach, J., Hansenne, M., Weytens, F., Dupuis, P., & Mikolajczak, M. 2011. Increasing emotional competence improves psychological and physical well-being, social relationships, and employability. *Emotion*, 11(2): 354–366. Doi: 10.1037/a0021554 (accessed July 22, 2020).

OECD. 2015. *Skills for Social Progress: The Power of Social and Emotional Skills*. Paris: OECD. http://www.oecd.org/education/skills-for-social-progress-9789264226159-en.htm (accessed 12 July 2020).

OECD. 2018. *The Future of Education and Skills - Education 2030*. Paris: OECD. https://www.oecd.org/education/2030/E2030%20Position%20Paper%20(05.04.2018).pdf (accessed 12 July 2020).

Payton, J., Weissberg, R. P., Durlak, J. A., Dymnicki, A. B., Taylor, R. D., & Schellinger, K. B., & Pachan, M. 2008. *The Positive Impact of Social and Emotional Learning for Kindergarten to Eighth-Grade Students: Findings from Three Scientific Reviews*. Chicago, IL: Collaborative for Academic, Social, and Emotional Learning. https://files.eric.ed.gov/fulltext/ED505370.pdf (accessed July 20, 2020).

Richardson, R. C., & Evans, E. T. 1997. Social and emotional competence: Motivating cultural responsive education. *Paper Presented at the Annual Conference and Exhibit of the Association for Supervision and Curriculum Development*. https://files.eric.ed.gov/fulltext/ED411944.pdf (accessed July 22, 2020).

Rubin, K. H., Bukowski, W.M., & Parker, J.G. (2006). Peer interactions, relationships, and groups. In *Handbook of Child Psychology: Social, Emotional, and Personality Development*, eds. N. Eisenberg, W. Damon, & R. M. Lerner, 571–645. Hoboken, NJ: Wiley.

Schutte, N. S., Malouff, J. M., Simunek, M., McKenley, J., & Hollander, S. 2002. Characteristic emotional intelligence and emotional well-being. *Cognition and Emotion*, 16(6): 769–785. Doi: 10.1080/02699930143000482 (accessed 10 July 2020).

Subhash, P. D., & Ashita R. 2019. *A study of Global Citizenship Education (GCED) in India*. Research Report. New Delhi: NCERT.

Student Police Cadet Directorate. 2017. "The SPC Project". https://studentpolicecadet.org/Home/spcprogram (accessed: 10 July 2020).

UNESCO. 1996. *Learning: The Treasure Within*. Report of UNESCO of the International Commission on Education for the Twenty-first Century. Paris: UNESCO.

UNESCO. 2014a. *Global Citizenship Education. Preparing Learners for the Challenges of the 21st Century*. Paris: UNESCO.

UNESCO. 2014b. *Teaching Respect for All: Implementation Guide. Part 3- Support Materials for Teaching and Learning: Guide for Educators*, 126–173. Paris: UNESCO. https://unesdoc.unesco.org/ark:/48223/pf0000227983 (accessed July 22, 2020).

UNESCO. 2015. *Global Citizenship Education: Topics and Learning Objectives*. Paris: UNESCO.

UNESCO. 2016. *Global Monitoring of Target 4.7: Themes in National Curriculum Frameworks*. Background paper prepared for the 2016 Global Education Monitoring Report, 37. Paris, UNESCO.

UNESCO 2016. *Education 2030: Towards Inclusive and Equitable Quality Education and Lifelong Learning for All* (Incheon Declaration and Framework for Action). Education 2030 Steering Committee. http://uis.unesco.org/sites/default/files/documents/education-2030-incheon-framework-for-action-implementation-of-sdg4-2016-en_2.pdf (accessed 12 July 2020).

United Nations. 2015. *Transforming Our World: The 2030 Agenda for Sustainable Development*. A/RES/70/1. United Nations.

Weissberg, R., Durlak, J., Domitrovich, C., & Gullotta, T. 2015. Social and emotional learning: Past, present, and future. In *Handbook of Social and Emotional Learning*, eds. J. A. Durlak, C. E. Domitrovich, R. P. Weissberg, & T. P. Gullota, 20–32. New York, NY: Guilford Publications.

Wolff, S. B., Pescosolido, A. T., & Druskat, V. U. 2002. Emotional intelligence as the basis of leadership emergence in self-managing teams. *Leadership Quarterly*. 13(5) (October): 505–510, Doi: 10.1016/S1048-9843(02)00141-8 (accessed July 22, 2020).

World Health Organization (WHO). 1997. *WHOQOL Measuring Quality of Life*. Geneva: WHO. https://www.who.int/mental_health/media/en/68.pdf (accessed 12 July 2020).

World Health Organization, (WHO). Global Health Estimates. 2017. *Depression and Other Common Mental Disorders*. Geneva: WHO, https://apps.who.int/iris/bitstream/handle/10665/254610/ -eng.pdf (accessed August 16, 2020).

Zins, J., Bloodworth, M., Weissberg, R., & Walberg, H. 2004. The scientific base linking social and emotional learning to school success. In *Building Academic Success on Social and Emotional Learning: What Does the Research Say?* eds. J. E. Zins, R. P. Weissberg, M. C. Wang, & H. J. Walberg, 3–22. New York: Teachers College Press.

6 Technological Innovation for Environmental Sustainability and Quality of Life

Luigi Aldieri and Concetto Paolo Vinci
University of Salerno

CONTENTS

6.1 INTRODUCTION

The technology has become more and more complex phenomenon, based on specific skills and important feature to improve the opportune advantages of the services and products, in relation to the tertiary sector in progress and strong coordination of the actions of firms (Rullani, 2009). These are the characteristics of the theoretical foundations evidenced in more investigations about innovation research topic. Knowledge derives from the realized network from one firm activity and the connections flows outside the firm (Mikucka et al., 2017; Iammarino, 2005; Quatraro, 2010).

During the last few years, there have been an increasing number of natural resources economic studies, with particular importance of environmental issues, good quality of life and connections with advanced social points (Guillen-Royo, 2019; Buchs and Koch, 2017; Glaeser and Resseger, 2010). From one perspective, there are factors based on intangible and localized benefits, the territorial capital, as discussed in Camagni (2009) and Farole et al. (2011). From another perspective, there are papers explaining the relevance of intangible weaknesses of context in poor areas even if with opportune system of financial or fiscal incentives (Dolan et al., 2008).

In order to deal with external shocks on economic effects, the institutional capital is assuming a key role, because all Government factors generating growth, such as environment, networks and knowledge, are relative to institutions' quality (Andersson et al., 2014; Dasgupta, 2005; Acemoglu, 2009).

Moreover, it has been evidenced that the social conditions can have an important impact on capital investment, the resources allocation process and the networks creation through long-run expectations (Castellacci and Tveito, 2018; Ravallion, 2010). In particular, Crescenzi and Rodríguez-Pose (2009) explore the social forces influencing the inequalities among the different geographical areas, in such a way that the competitiveness of industrial systems of innovation can be supported. Indeed, Royuela et al. (2011) discussed that the life quality in large geographical areas can produce important repercussions on the growth perspectives of each industrial context.

The aim of this chapter is develop a theoretical foundation where the institutional, social and life quality inputs are able to explain the transformation of the technology into new outside knowledge at the firm level. We consider the evolutionary approach to innovation process: in this perspective, institutional and social context determine relevant technological competences from the interaction between different economic agents (Binder and Blankenberg, 2017; Von Tunzelmann and Wang, 2007). Hence, the socio-institutional conditions affect the knowledge absorption capacity and the economic growth of firms. These features allow the persistence of inequalities in the different industrial sectors.

6.2 LITERATURE REVIEW

The productions of services and goods, and government actions have the function of improving the quality of life of people. In economics, the quality of life can be proxied by happiness or utility, while we get subjective well-being (SWB) in psychology. Moreover, economists usually assume that income can be employed to measure utility (Clark et al., 2008a; Corral-Verdugo et al., 2011; Delle Fave et al., 2011; Ryff, 1989). In the economic theory and models, own consumption of goods and services is the main variable of the agents' utilities. According to these models, pursuit of individual self-interest improves the quality of life and then the happiness (Koo et al., 2019; Hellevik, 2015; Korpela et al., 2014; Gasper, 2005). Happiness is evaluated through some measures of income (Veenhoven, 1993; Helliwell et al., 2017; Clark and Oswald, 1994; Conti and Pudney, 2011). Economic policies actions favouring the social welfare against the poverty assume a key role in economic growth theory.

In psychology perspective, the academicians prefer to directly measure SWB by implementing a large sample survey (Benjamin et al., 2014; Chadi, 2019; Abowd and Stinson, 2013; Angelini et al., 2017; Clark et al., 2008b; Baetschmann, 2014). Indeed, life quality is usually measured through a scale to evaluate the satisfaction degree where the question is the following: "All things considered, how satisfied are you with your life as a whole these days?"

There is a contrast between standard economic theory on welfare economics and empirical evidence concerning the studies on SWB survey data. Indeed, the findings suggest a fundamental conflict between social and individual quality of life, a true paradox: most people wish high income, but even if societies have become richer, they do not become happier. According to Graham (2005):

> While most happiness studies find that within countries wealthier people are, on average, happier than poor ones, studies across countries and over time find very little, if any,

relationship between increases in per capita income and average happiness levels. On average, wealthier countries (as a group) are happier than poor ones (as a group); happiness seems to rise with income up to a point, but not beyond it. Yet even among the less happy, poorer countries, there is not a clear relationship between average income and average happiness levels, suggesting that many other factors – including cultural traits – are at play.

This phenomenon is defined as the "Easterlin paradox" (Easterlin et al., 2010). This paradox is verified in Japan, the United States and Europe. For this reason, there is an increasing economic discussion, according to which the quality of life and happiness instead of income should become the main task for policymakers. Indeed, they should use the happiness index measures rather than the gross domestic product (GDP) to account for the welfare of countries (Fleurbaey, 2009; Frey and Stutzer, 2012; Greco et al., 2020; Contoyannis et al., 2004; Mangaraj and Aparajita, 2020).

On the basis of the recent contributions on life quality, many academicians wonder whether utility or the standard economic theory can be considered yet consistent both theoretically and empirically (Engelbrecht, 2018; Goebel et al., 2015; Heffetz and Robin, 2013; Heffetz and Reeves, 2019). Many economists and psychologists concluded that only the subjective approach is used to model human quality of life, rather than objective one. Numerous studies have explored welfare levels in economic theories. Indeed, some scholars believe that "the pursuit of individual self-interest is not a good formula for personal happiness" (Layard, 2005). Since this stream of research is considered as non-mainstream, most economists have ignored it and there are too few formal models on people life quality.

As discussed in the literature (Aldieri et al., 2019, 2020; Kyle, 2020; Kassenboehmer and Haisken-De New, 2012; Meyer and Sullivan, 2003), the exploration of over-estimated and under-estimated effects of happiness on growth deserves further observation and discussion. To this end, we will implement a theoretical economic model to explain and solve the Easterlin Paradox.

On the one hand, relative to empirical and experimental estimations, individuals should care about important factors like family life, friendship and health. Indeed, these factors affect people's behaviour towards economic actions (Schubert, 2015; Hirvillami et al., 2013; Graham and Nikolova, 2013; Hyll and Schneider, 2013; Pénard et al., 2013; Rözer and Kraaykamp, 2013; Howell et al., 2011; Graham and Pettinato, 2002). As explained in Diener and Seligman (2004), there are several studies that evidence the impact of non-income elements. As discussed in Di Tella and MacCulloch (2006), this "omitted variables" procedure could deepen the Easterlin paradox. Additionally, the "omitted variables" procedure represents an instrument to overcome the paradox (Sabatini and Saraceno, 2017; Kavetsos and Koutroumpis, 2011; Stepanikova et al., 2010; Alesina et al., 2004; Di Tella et al., 2001).

The other approach, Easterlin (1995, 2001) pays attention to the income itself and predicts that quality of life is generated on the basis of the difference between some aspiration level and income. Indeed, the aspiration theory is an alternative of social comparison procedure. This stream of research seems to evidence that non-material variables can fundamentally improve the people life quality (Frijters and Beatton, 2012; Stiglitz et al., 2010; Campanelli and O'muircheartaigh, 2002). In this chapter, we explain that the non-income variables might be relevant because they generate a positive spillover effect outside the economic context of single agent.

In particular, other theoretical and economic models are needed to explore individuals' welfare, especially from the perspective of Pareto efficiency condition. On the basis of our knowledge, there are few studies that consider efficient choice problems such as social happiness maximization and individual optimal choice (Ng and Wang, 1993; Chadi, 2015; Ng and Ng, 2001; Ng, 2003).

There are studies that specifically identify a critical income level to evidence the impact of happiness (Kasser, 2017; Van Praag, 2011; Van Praag et al., 2011; Brockmann et al., 2009; Bruni and Stanca, 2006, 2008). However, they do not consider the Pareto efficiency condition. Hockman and Rodgers (1969) believe in interdependent preferences, which can lead to redistributive activities. Boskin and Sheshinski (1978) investigated the efficient taxation process in case people welfare is associated to the relative income. We will use a basic *Non-Overlapping-Growth model (NOG)* to model the generation of social externalities in the impact of innovation on quality of life through environmental sustainability perspective (Ericson et al., 2014).

6.3 THEORETICAL UNDERPINNING

Following Acemoglu (1996) and Aldieri and Vinci (2017), we build our model on a basic *NOG* framework, where a continuum of entrepreneurs and employees are normalized to unity and breath for two phases. In the first period of their life, they invest in human capital while tycoons decide on the optimal levels of *R&D* capital[1]; in the following stage at the beginning, manufacture will occur in a sort of one to one joint-venture of a company and a worker, then agents spend all their money and die leaving no endowments.

Employees take their decisions concerning human capital with no idea which entrepreneur they will match up, so their choices be governed by their expected *R&D* capital levels. The reverse is also true, in the sense that companies' decisions will depend on the expected human capital. The match is assumed to be one to one and unchangeable once formed because of high costs. It's our opinion that a safety, happiness, health condition and social environment, in a few words, are a quality of life indicator due to the Government ruling activity. In this light, we consider the following production function:

$$Y_{i,j} = A h_i{}^{\alpha} k_j{}^{(1-\alpha)} \tag{6.1}$$

where K_j and h_i capture, respectively, the j firm *R&D* capital levels *and human capital of the i* employee.

A is a positive parameter capturing both the above exogenous qualitative components and other factors (the state of technology) influencing production. In line with search-models literature, we assume that a bargaining process, between workers and companies, run the income-sharing regime and defines the share [β for worker and $(1-\beta)$ for the firm] of the total production.

The assumed matching function's randomness implies that each firm (worker), having no idea of the counterpart, considers her/his expected return depending on

---

[1] It's only for simplicity that we don't consider physical capital within production.

the whole distribution of human (R&D) capital across all the employees (entrepreneurs). The expected values of total wage and firms' income may be written as:

$$W_i = \beta A h_i^{\alpha} \int_0^1 K_j^{(1-\alpha)} \, dj \qquad (6.2)$$

$$R_j = (1-\beta) K_j^{(1-\alpha)} \int_0^1 h_i^{\alpha} \, di \qquad (6.3)$$

The utility function of the i worker is:

$$U_i = C_i - \frac{\theta_i \, h_i^{(1+\gamma)}}{1+\gamma} \qquad (6.4)$$

with C_i standing for consumption, θ_i, whose distribution across employment is assumed to be common knowledge, is a taste positive parameter measuring disutility from investment in human capital. Equation (6.4) has to be maximized with respect to the budget constraint:

$$C_i \leq W \qquad (6.5)$$

The *f.o.c.* of the above-constrained maximization will give us the following:

$$h_i = \left[\frac{\beta \alpha A \int_0^1 K_j^{(1-\alpha)} \, dj}{\theta_i} \right]^{\frac{1}{(1+\gamma-\alpha)}} \qquad (6.6)$$

On the companies' side, the objective will be that of investing in *R&D* capital in order to maximize the expected profits given by:

$$\Pi_{i,j} = R_j - cuK_j \qquad (6.7)^2$$

from inspection of which we derive

$$K_j = \left[\frac{(1-\beta)(1-\alpha) A \int_0^1 h_i^{\alpha} \, di}{cu} \right]^{\frac{1}{\alpha}} \qquad (6.8)$$

From the above, we may derive what follows:

---

[2] It's obviously assumed unity for output price, while cu is the user cost of R&D capital.

Proposition 6.1

Assuming that $\theta_i = \theta$, a random matching function and the above income-sharing regime:

1. There exists a unique equilibrium in the decentralized search economy.
2. The above equilibrium is Pareto-inefficient and displays social increasing returns in a way that tiny surges in R&D make all wealthier.

Proof of Proposition 6.1
The equilibrium values derive by combining equations (6.6) and (6.8)

Proof of Proposition 6.2
To prove Pareto-inefficiency, we take firms' profits in the following way:

$$\Pi = (1-\beta)K^{*(1-\alpha)}h^{*\alpha} - cuK^*$$

and determine outcomes due to tiny variations in the equilibrium values K^*, h^*. We obtain:

$$d\Pi = \left[(1-\beta)K^{*(1-\alpha)}h^{*(\alpha-1)}\right]dh^* + \left[(1-\beta)K^{*(-\alpha)}h^{*\alpha} - cu\right]dK^*$$

from which we may observe as terms multiplied by dK^* and dh^* are zero by f.o.c., so $d\Pi = 0$.

Proposition 6.2

Once a band of workers invests more in human capital, employers answer, hence other workers' equilibrium revenues will get better.

Proof of Proposition 6.2
Following Acemoglu (1996) if a quota m of employees experiment a cut of θ from θ_1 to θ_2 equation (6.8) will turn into:

$$h_1 = \left[\frac{\beta\alpha A \int_0^1 K_j^{(1-\alpha)}dj}{\theta_1}\right]^{\frac{1}{(1+\gamma-\alpha)}} \tag{6.9}$$

$$h_2 = \left[\frac{\beta\alpha A \int_0^1 K_j^{(1-\alpha)}dj}{\theta_2}\right]^{\frac{1}{(1+\gamma-\alpha)}} \tag{6.10}$$

With simple algebraic manipulations, we may derive:

$$K_j = \left\{\frac{(1-\beta)(1-\alpha)A\left[mh_2^{\alpha} + (1-m)h_1^{\alpha}\right]}{cu}\right\}^{\frac{1}{\alpha}} \tag{6.11}$$

from which we may easily derive that R&D capital to human capital ratio of workers who have not experimented a cut in θ increase with m.

The above propositions put in evidence the role of social increasing returns in R&D and human capital. We have showed as if, for example, a small group of workers decide to invest more in human capital, entrepreneurs will answer by increasing investments in R&D capital, and as a consequence, the earnings of both employers and employees will get better. The reverse is obviously true, in the sense that if employers invest more in innovations by using more R&D capital, workers answer by investing more in human capital. In this way, a virtuous circle will start with continuous benefits for the whole community. A really quite noteworthy policy implication concerns those programmes aimed at improve health, safety conditions, social environment and quality of life in general. We can conclude as an economic policy of *Good Governance* makes people happier and this happiness act as a fly for the above virtuous circle guaranteed by the presence of pecuniary externalities.

6.4 DISCUSSION

This chapter develops a theoretical model that describes the foundation for analysing quality of life from the individual perspectives. We unify the "omitted variables" approach and aspiration theory (Frank, 1985, 1997; Easterlin, 2003). In particular, the association between happiness and income can be explored through the mainstream economics framework (Luttmer, 2005; Neumark and Postlewaite, 1993; Frank and Sunstein, 2001; Solnick and Hemenway, 2006).

In the model, we assume that a continuum of entrepreneurs and employees are normalized to unity and breath for two phases. In the first period of their life, they invest in human capital, while tycoons decide on the optimal levels of *R&D* capital; in the following stage at the beginning, manufacture will occur in a sort of one to one joint-venture of a company and a worker, then agents spend all their money and die leaving no endowments.

Employees take their decisions concerning human capital with no idea which entrepreneur they will match up, so their expected R&D capital levels govern their choices. The reverse is also true, in the sense that companies' decisions will depend on expected human capital. The match is assumed to be one to one and unchangeable once formed because of high costs. It's our opinion that a safety, happiness, health condition and social environment, in a few words, are a quality of life indicator due to the Government ruling activity. We consider a positive parameter capturing both the above exogenous qualitative components and other factors (the state of technology) influencing production. In line with search-models literature, we assume that a bargaining process, between workers and companies, run the income-sharing regime and defines the share of the total production.

The assumed matching function's randomness implies that each firm (worker), having no idea of the counterpart, considers her/his expected return depending on the whole distribution of human (R&D) capital across all the employees (entrepreneurs). After opportune maximization procedure, we get that the derived equilibrium is Pareto-inefficient and displays social increasing returns in a way that tiny surges in

R&D make all wealthier. Moreover, once a band of workers invests more in human capital, employer's answer, hence other workers' equilibrium revenues will get better.

Therefore, social spillovers are responsible for a contribution to happiness, through non-income variables.

The significance and variety of the findings are peculiar for policy implications: both income variable and non-income variables are relevant to improve happiness. Indeed, this prediction can be observed specifically in our model. Thus, to avoid a reduction in the happiness levels, the public expenditures for welfare should be financed, so basic mental status, fighting unemployment, human rights, health, inflation and family life can be supported.

These findings are very recent in the literature (Usai et al., 2020; Zhao and Sun, 2020). Indeed, according to Easterlin (2005) "the cross sectional relationship is not necessarily a trustworthy guide to experience over time or to inferences about policy", and evidenced that there is zero marginal utility of income rather than diminishing marginal utility of income in both among-country analysis and within-country one.

6.5 LIMITATIONS AND CONCLUSIONS

The results obtained in this chapter evidence that the self-interested behaviour and optimization approach conditions should be employed in case of happiness economics exploration. The psychological approach should consider the individual rational choice and social welfare maximization. The economic approach should consider also non-material variables, as discussed in the happiness economics literature.

However, the analysis suffers from some weaknesses. In fact, the impact of spillovers in the innovation process could differ due to the technological proximity between firms or due to geographical distance. To this end, it would be appropriate to develop an analysis based on the use of data on patents of companies classified by technological fields in order to compute a symmetric or asymmetric index according to the way we intend to take into account the different bargaining power on the market.

Moreover, the model developed does not consider how the different firms can exploit and apply the external innovation. This feature is relative to the absorptive capacity component, here not explored in the theoretical framework.

On the basis of the previous points highlighted, future research can be implemented.

REFERENCES

Abowd, J. M. and M. H. Stinson. 2013. Estimating measurement error in annual job eranings: A comparison of survey and administrative data. *Review of Economics and Statistics,* 95: 1451–1467.

Acemoglu, D. 1996. A microfoundation for social increasing returns in human capital accumulation. *The Quarterly Journal of Economics,* 111: 779–804.

Acemoglu, D. 2009. *Introduction to Modern Economic Growth.* Princeton University Press, Princeton, NJ.

Aldieri, L., B. Bruno, and C. P. Vinci. 2019. Does environmental innovation make us happy? An empirical investigation. *Socio-economic Planning Sciences,* 67: 166–172.

Aldieri, L., B. Bruno, and C. P. Vinci. 2020. A multi-dimensional approach to happiness and innovation. *Applied Economics,* Doi: 10.1080/00036846.2020.1828807.

Aldieri, L. and C. P. Vinci. 2017. R&D spillovers and productivity in Italian manufacturing firms. *International Journal of Innovation Management,* Doi: 10.1142/S1363919617500359.

Alesina, A., R. Di Tella, and R. MacCulloch. 2004. Inequality and happiness: Are Europeans and Americans different? *Journal of Public Economics,* 88: 2009–2042.

Andersson, D., J. Nassen, J. Larsson, and J. Holmberg. 2014. Greenhouse gas emissions and subjective wee-being: An analysis of Swedish households. *Ecological Economics,* 102: 75–82.

Angelini, V., M. Bertoni, and L. Corazzini. 2017. Unpacking the determinants of life satisfaction: A survey experiment. *Journal of the Royal Statistical Society Statistics in Society Series A,* 180: 225–246.

Baetschmann, G. 2014. Heterogeneity in the relationship between happiness and age: Evidence from the German Socio-economic panel. *German Economic Review,* 15: 393–410.

Benjamin D. J., O. Heffetz, M. S. Kimball, and A. Rees-Jones. 2014. Can marginal rates of substitution be inferred from happiness data? Evidence from residency choices. *American Economic Review,* 104: 3498–3528.

Binder, M. and A.-K. Blankenberg. 2017. Green lifestyles and subjective weel-being: More about self-image than actual behavior? *Journal of Economic Behavior and Organization,* 137: 304–323.

Boskin, M. J. and E. Sheshinski. 1978. Optimal redistributive taxation when individual welfare depends upon relative income. *Quarterly Journal of Economics,* 92: 589–601.

Brockmann, H., J. Delhey, C. Welzel, and H. Yuan. 2009. The China puzzle: Falling happiness in a rising economy, *Journal of Happiness Studies,* 10: 387–405.

Bruni, L. and L. Stanca. 2006. Income aspirations, television and happiness: Evidence from the world values survey. *Kyklos,* 59: 209–225.

Bruni, L. and L. Stanca. 2008, Watching alone: Relational goods, television and happiness. *Journal of Economic Behavior & Organization,* 65: 506–528.

Buchs, M. and M. Koch. 2017. *Postgrowth and Wellbeing.* Palgrave Macmillan, London.

Camagni R. 2009. Territorial capital and regional development. In *Handbook of Regional Growth and Development Theories,* Capello R., P. Nijkamp, and Cannari L. (eds.), Elsevier, Cheltenham *Mezzogiorno e politiche regionali,* Banca d'Italia, Roma.

Campanelli P. and C. O'muircheartaigh. 2002. The importance of experimental control in testing the impact of interviewer continuity on panel survey nonresponse. *Quality and Quantity,* 36: 129–144.

Castellacci, F. and V. Tveito. 2018. Internet use and well-being: A survey and a theoretical framework. *Research Policy,* 47: 308–325.

Chadi, A. 2015. Concerns about the Euro and happiness in Germany during times of crisis. *European Journal of Political Economy,* 40: 126–146.

Chadi, A. 2019. Dissatisfied with life or with being interviewed? Happiness and the motivation to partcipate in a survey. *Social Choice and Welfare,* 53: 519–553.

Clark, A. E., E. Diener, Y. Georgellis, and R. E. Lucas. 2008a. Lags and leads in life satisfaction: A test of the baseline hypothesis. *Economic Journal,* 118:F222–F243.

Clark, A. E., P. Frijters, and M. A. Shields. 2008b. Relative income, happiness, and utility. *Journal of Economic Literature,* 95: 115–122.

Clark, A. E. and A. J. Oswald. 1994. Unhappiness and unemployment. *Economic Journal,* 104: 648–659.

Conti G. and S. Pudney. 2011. Survey design and the analysis of satisfaction. *Review of Economics and Statistics,* 93: 1087–1093.

Contoyannis P., A. M. Jones, and N. Rice. 2004. The dynamics of health in the British household panel survey. *Journal of Applied Economics,* 19: 473–503.

Corral-Verdugo, V., J. Mireles-Adcosta, C. Tapia-Fonliem, and B. Fraijo-Sing. 2011. Happiness as correlate of sustainable behavior: A study of pro-ecological, frugal, equitable and ultruistic actions that promote subjective well-being. *Human Ecology Review,* 18: 95–104.

Crescenzi, R. and A. Rodrìguez-Pose. 2009. Systems of innovation and regional growth in the EU: Endogenous vs. external innovative activities and socio-economic conditions. In Fratesi U. and Senn L. (eds.), *Growth and Innovation of Competitive Regions: The Role of Internal and External Connections,* Advances in spatial science, Berlin: Springer-Verlag.

Dasgupta, P. 2005. The economics of social capital. *The Economic Record,* 81 (Supplement): 2–21.

Delle Fave, A., I. Brdar, D. Vella-Brodrick, and M. P. Wissing. 2011. The eudaimonic and hedonic components of happiness: Qualitative and quantitative findings. *Social Indicator Research,* 100: 185–207.

Diener, E., and M. Seligman. 2004. Beyond money: Toward an economy of well-being. *Psychological Science in the Public Interest,* 5: 1–31.

Di Tella, R. and R. J. MacCulloch. 2006. Some uses of happiness data in economics. *Journal of Economic Perspectives,* 20: 25–46.

Di Tella, R., R. MacCulloch, and A. Oswald. 2001. Preferences over inflation and unemployment. Evidence from surveys of happiness. *The American Economic Review,* 91: 335–341.

Dolan, P., T. Peasgood, and M. White. 2008. Do we really know what makes us happy? A review of the economic literature on the factors associated with subjective well-being. *Journal of Economic Psychology,* 29: 94–122.

Easterlin, R. A. 1995. Will raising the incomes of all increase the happiness of all? *Journal of Economic Behavior and Organization,* 27: 35–47.

Easterlin, R. A. 2001. Income and happiness: Toward a unified theory. *Economic Journal,* 111: 465–484.

Easterlin, R. A. 2003. Explaining happiness. *Proceedings of the National Academy of Sciences,* 100: 11176–11183.

Easterlin, R. A. 2005. Diminishing marginal utility of income? A caveat emptor. *Social Indicators Research,* 70: 243–255.

Easterlin, R. A., L. A. McVey, M. Switek, O. Sawangfa, and J. S. Zweig. 2010. The happiness–income paradox revisited. *Proceedings of the National Academy of Sciences,* 107: 22463–22468.

Engelbrecht, H. J. 2018. The (social) innovation–subjective well-being nexus: Subjective well-being impacts as an additional assessment metric of technological and social innovations. Innovation. *The European Journal of Social Science Research,* 31: 317–332.

Ericson, T., B. G. Kjonstad, and A. Barstad. 2014. Mindfulness and sustainability. *Ecological Economics,* 104: 73–79.

Farole, T., A. Rodrìguez-Pose, and M. Storper. 2011. Cohesion policy in the European Union: Growth, geography and institutions. *Journal of Common Market Studies,* 5: 1089–1111.

Fleurbaey, M. 2009. Beyond GDP: The quest for a measure of social welfare. *Journal of Economic Literature,* 47: 1029–1075.

Frank, R. H. 1985. *Choosing the Right Pond: Human Behavior and the Quest for Status.* Oxford University Press, New York.

Frank, R. H. 1997. The frame of reference as a public goods. *Economic Journal,* 107: 1832–1847.

Frank, R. H., and C. R. Sunstein. 2001. Cost-benefit analysis and relative position. *University of Chicargo Law Review,* 68: 323–375.

Frey, B. S., and A. Stutzer. 2012. The use of happiness research for public policy. *Social Choice and Welfare,* 38: 659–674.

Frijters, P., and T. Beatton. 2012. The mystery of the U-shaped relationship between happiness and age. *Journal of Economic Behavior & Organization*, 82: 525–542.

Gasper, D. 2005. Subjective and objective well-being in relation to economic inputs: Puzzles and responses. *Review of Social Economy*, 63: 177–206.

Glaeser, E., and M. Resseger. 2010. The complementarity between cities and skills. *Journal of Regional Science*, 50 (1): 221–244.

Goebel J., C. Krekel, T. Tiefenbach, and N. R. Ziebarth. 2015. Natural disaster, policy action, and mental well-being: The case of how natural disasters can affect environmental concerns, risk aversion, and even politics: Evidence from Fukushima and three European countries. *Journal of Population Economics*, 28: 1137–1180.

Graham, C. 2005. The economics of happiness, in S. Durlauf and L. Blume, eds., *The New Palgrave Dictionary of Economics*, 2nd Edition. Palgrave Macmillan UK.

Graham, C. and M. Nikolova. 2013. Does access to information technology make people happier? Insights from well-being surveys from around the world. *The Journal of Socio-Economics*, 44: 126–139.

Graham, C. and S. Pettinato. 2002. Frustrated achievers: Winners, losers and subjective well-being in new market economies. *Journal of Development Studies*, 38: 100–140.

Greco, S., A. Ishizaka, G. Resce, and G. Torrisi. 2020. Measuring well-being by a multi-dimensional spatial model in OECD better life index framework. *Socio-Economic Planning Sciences*, Doi: 10.1016/j.seps.2019.01.006.

Guillen-Royo, M. 2019. Sustainable consumption and wellbeing: Does on-line shopping matter? *Journal of Cleaner Production*, 229: 1112–1124.

Heffetz, O. and M. Rabin. 2013. Conclusions regarding cross-group differences in happiness depend on difficulty of reaching respondents. *American Economic Review*, 103: 3001–3021.

Heffetz, O. and D. B. Reeves. 2019. Difficulty to reach respondents and nonresponse bias: Evidence from large government surveys. *Review of Economics and Statistics*, 101: 176–191.

Hellevik, O. 2015. Is the good life sustainable? A three-decade study of values, happiness and sustainability in Norway. In Lykke Syse, K. and Mueller, M. L. (eds.), *Sustainable Consumption and the Good Life*, Routledge, Abingdon.

Helliwell, J., R. Layard, and J. Sachs. 2017. *World Happiness Report 2017*. Sustainable Development Solutions Network, New York.

Hirvillami, T., S. Laakso, M. Lettenmeier, and S. Lahteenoja. 2013. Studying weel-being and its environmental impacts: A case study of minimum income receivers in Finland. *Journal of Human Development and Capabilities*, 14: 134–154.

Hockman, H. M. and J. D. Rodgers. 1969. Pareto optimal redistribution. *American Economic Review*, 59: 542–557.

Howell, A. J., R. L. Dopko, H.-A. Passmore, and K. Buro. 2011. Nature connectedness: Associations with well-being and mildfulness. *Personality and Individual Differences*, 51, 166–171.

Hyll, W. and L. Schneider. 2013. The causal effect of watching TV on material aspirations: Evidence from the valley of the innocent. *Journal of Economic Behavior and Organization*, 86: 37–51.

Iammarino, S. 2005. An evolutionary integrated view of regional systems of innovation: Concepts, measures and historical perspectives. *European Planning Studies*, 13: 495–517.

Kassenboehmer, S. C. and J. P. Haisken-DeNew. 2012. Heresy or enlightenment? The well-being age u–shape effect is flat. *Economics Letters*, 117:235–238.

Kasser, T. 2017. Living both well and sustainably: A review of the literature, with some reflections on future research, interventions and policy. *Philosophical Transactions of the Royal Society A*, Doi: 10.1098/rsta.2016.0369.

Kavetsos, G. and P. Koutroumpis. 2011. Technological affluence and subjective well-being. *Journal of economic psychology*, 32: 742–753.

Koo, J., Y. J. Choi, and I. Park. 2019. Innovation and welfare: The marriage of an unlikely couple. *Policy and Society,* 39: 189–207.

Korpela, K., K. Borodulin, M. Neuvonen, O. Paronen, O., and L. Tyrvainen. 2014. Analyzing the mediators between nature-based outdoor recreation and emotional well-being, *Journal of Environmental Psychology*, 37: 1–7.

Kyle, M. K. 2020. The alignment of innovation policy and social welfare: Evidence from pharmaceuticals. *Innovation Policy and the Economy,* Doi: 10.1086/705640.

Layard, R. 2005. *Happiness.* London: Penguin Press.

Luttmer, E. F. P. 2005. Neighbors as negatives: Relative earnings and well-being. *Quarterly Journal of Economics*, 120: 963–1002.

Mangaraj, B. K. and U. Aparajita. 2020. Constructing a generalized model of the human development index. *Socio-economic Planning Sciences,* Doi: 10.1016/j.seps.2019.100778.

Meyer B. D. and J. X. Sullivan. 2003. Measuring the wellbeing of the poor using income and consumption. *The Journal of Human Resources,* 38: 1180–1220.

Mikucka, M., F. Sarracino, and J. K. Dubrow. 2017. When does economic growth improve life satisfaction? Multilevel analysis of the roles of social trust and income inequality in 46 countries, 1981–2012. *World Development*, 93: 447–459.

Neumark, D. and A. Postlewaite. 1998. Relative income concerns and the rise in maried women's employment. *Journal of Public Economics*, 70: 157–83.

Ng, Y.-K. 2003. From preference to happiness: Towards a more complete welfare economics. *Social Choice and Welfare*, 20: 307–350.

Ng, Y.-K. and J. Wang. 1993. Relative income, aspiration, environmental quality, individual and political myopia: Why may the rat-race for material growth be welfare-reducing? *Mathematical Social Sciences*, 26: 3–23.

Ng, S. and N. Yew-Kuang. 2001. Welfare-reducing growth despite individual and government opitimization. *Social Choice and Welfare*, 18: 497–506.

Pénard, T., N., Poussing, and R. Suire. 2013. Does the internet make people happier? *The Journal of Socio-Economics*, 46: 105–116.

Quatraro, F. 2010. Knowledge coherence, variety and producyivity growth: Manufacturing evidence from Italian regions. *Research Policy*, 39: 1289–1302.

Ravallion, M. 2010. Do poorer countries have less capacity for redistribution? *Journal of Globalization and Development*, 1(2): 1–29.

Royuela, V., R. Moreno R., and E. Vaya. 2010. Influence of quality of life on urban growth: A case study of Barcelona, Spain. *Regional Studies*, 44(5): 551–567.

Rözer, J. and G. Kraaykamp. 2013. Income inequality and subjective well-being: A cross-national study on the conditional effects of individual and national characteristics. *Social Indicators Research*, 113: 1009–1023.

Rullani, E. 2009. Knowledge economy and local development: The evolution of industrial districts and the new role of 'urban networks'. *Review of Economic Conditions in Italy*, 2: 237–284.

Ryff, C. D. 1989. Happiness is everything, or is it? Explorations on the maning of psychological well-being. *Journal of Personality and Social Psychology,* 57: 1069–1081.

Sabatini, F. and F. Saraceno. 2017. Online networks and subjective well-being. *Kyklos*, 70: 456–480.

Schubert, C. 2015. What do we mean when we say that innovation and entrepreneurship (policy) increase "welfare"? *Journal of Economic Issues*, 49: 1–22.

Solnick, S. J. and D. Hemenway. 2006. Are positional concerns stronger in some domains than in others? *American Economic Review, Papers and Proceedings*, 95: 147–151.

Stepanikova, I., N. H. Nie, and X. He. 2010. Time on the internet at home, loneliness, and life satisfaction: Evidence from panel time-diary data. *Computers in Human Behavior*, 26: 329–338.

Stiglitz, J. E., A. Sen, and J.-P Fitoussi. 2010. *Mis-Measuring Our Lives. Why the GDP Doesn't Add up.* New York: New Books.

Usai, A., B. Orlando, and A. Mazzoleni, A. 2020. Happiness as a driver of entrepreneurial initiative and innovation capital. *Journal of Intellectual Capital,* Doi: 10.1108/JIC-11-2019–0250.

Van Praag, B. 2011. Well-being inequality and reference groups: An agenda for new research. *The Journal of Economic Inequality*, 9: 111–127.

Van Praag, B. M. S., and A. Ferrer-i-Carbonell, A. 2011. Happiness economics: A new road to measuring and comparing happiness. *Foundations and Trends® in Microeconomics*, 6: 1–97. Doi: 10.1561/0700000026.

Veenhoven, R. 1993. *Happiness in Nations: Subjective Application of Life in 56 Nations.* Rotterdam Institute for Social Policy Research. Rotterdam: Erasmus University.

Von Tunzelmann N. and Q. Wang 2007. Capabilities and production theory. *Structural Change and Economic Dynamics*, 18: 192–211.

Zhao, X. and Z. Sun. 2020. The effect of satisfaction with environmental performance on subjective well-being in China: GDP as a moderating factor. *Sustainability.* Doi: 10.3390/su12051745.

7 Key Factors and Influencers Determining Work-Life Balance and Their Implications on Quality of Life

John Bourke
The Business Excellence Institute

Ole Petter Anfinsen
Anfinsen Executive Health & Performance

James Baker
Henley Business School

CONTENTS

7.1   INTRODUCTION

Finding the right work-life balance is highly important for one's quality of life. Given the increasing pressure and demands of twenty-first century living, it is of critical interest to both individuals and organizations.

Work is a part of life and we should not have to *live to work* or *work to live* but rather *live while (sometimes) working*. However, although some researchers such as Grady and McCarthy (2008) frame things in terms of *work-life integration*, people tend to see work and life as being discrete, polarized constructs as is implied by the very notion that work and life need to be balanced. They frame things as a zero-sum game in which they must continuously juggle the conflicting demands of work and private life. This has implications for our well-being and quality of life that are of concern for us as individuals, as teams, and as organizations.

Many people think of work-life balance as being about the conflicting demands on our time. However, some researchers – such as Greenhaus and Beutell (1985) and Carlson, Michele Kacmar, and Williams (2000) – view the conflict as being strain-based and behaviour-based as well as time-based and there is little consensus about the meaning of balance (Casper et al. 2018). So, what exactly is work-life balance and what are the key factors that determine it? This is what we explore in this chapter, investigating the phenomenon by combining the results of three independent, qualitative studies which build on a significant body of research, the large majority of which has been quantitative (Chang et al. 2010).

The authors hold that reality is layered, not flat, with a different ontology being required for the physical and the social worlds. Consequently, all three studies employed a subjectivist ontology and adopted an "engaged constructionist" approach (Easterby-Smith, Thorpe, and Jackson 2015) since phenomena such as work-life balance are subjective and, as such, the lens used to research them should be an interpretive one. The studies combined the perspective of an "empathic observer" with that of a "faithful reporter" (Blaikie 2009) as they wished to investigate the lived experiences of the participants – pursuing Weber's *verstehen* (Crotty 1998) – while also allowing the participants to speak for themselves.

In contrast to the majority of work-life balance research which focuses on younger people (Emslie and Hunt 2009), in order to expand on the understanding of the phenomenon, the studies focused on people who are mid-career with plenty of experience pursuing, creating, and sustaining balance in their lives.

The first study (Study A) explored the experiences of people in a range of professions in Ireland. The second study (Study B) focused on the experiences of managers working in medium-sized UK companies, while the last study (Study C) explored the phenomenon as it is perceived by senior human resource managers in a multinational commercial organization in the UK. The different context of each of the studies provided the opportunity for triangulation and helped establish a contribution to knowledge since if something is to be considered key, it will most likely be a common factor across people in different contexts and since most work-life balance research has been conducted only in a single country (Thilagavathy and Geetha 2020).

Each of the authors was drawn to their research for different reasons. One because he has spent most of his career working with what many would consider to be no work-life balance and 15 years of it living in Japan where there is an official term for "death by overwork" (*karoushi*). One because he is passionate about executive health and well-being, and one because he is interested in the importance of work-life balance for recruiting. Each sought to investigate the key factors that determine work-life balance within the context of their study.

Triangulation of the findings of the three parallel studies provides insight into what the key factors might be in broader contexts. However, it is important to note that research is qualitative and its results relate to the contexts in which the studies were conducted and, in combination, to "English speaking Europe at the start of the 2020s." Consequently, care should be taken about generalizing the findings to a global audience, especially as the work-life balance construct is originally a Western one (Lewis and Beauregard 2018). Further research, both qualitative and quantitative, including longitudinal, studies to investigate if the key factors are stable over time would be useful to explore if and how the findings apply more broadly.

7.2 THE STUDIES

Two of three studies explored the phenomenon of work-life balance from an individual perspective, while the third did so from an organizational one. In keeping with the research philosophy outlined above, the studies were inductive in nature, enabling the phenomenon to be explored without being anchored by what we already know prior to being integrated with other knowledge (Gioia, Corley, and Hamilton 2013). As the studies were interpretive and relativist in approach, a large number of participants was not necessary to produce viable results. For example, when conducting interviews, a wide range of numbers is considered acceptable and a very small sample, with a lower limit at four, is deemed sufficient when taken from a relatively homogenous population (Saunders and Townsend 2016). The participant numbers in the studies were based on this principle, as well as on the nature of the research question and the research design.

All three studies used purposive sampling with participants being selected from the researchers' individual professional networks. This is a form of convenience sampling, which has been found by meta-analysis to be common in work-life balance studies (Chang et al. 2010).

Data were analysed by coding, an analytical process that enables themes to emerge from the data, and thematic analysis (for those interested, the coding methods were eclectic and involved structural, *in vivo*, initial, descriptive, attribute and values coding being used prior to second-cycle axial coding and analysis of emerging themes).

7.2.1 STUDY A

This study was conducted in Ireland. Within that context, in order to identify as wide a range of lived experiences as possible by obtaining input from people with different perspectives (Easterby-Smith, Thorpe, and Jackson 2015), the study was designed to keep participant selection as diverse as possible. Participants were selected from different domains of work and family structures as well as having different genders (see Table 7.1).

TABLE 7.1

Study A – Participant Attributes

Occupation	Gender	Personal Status	Age Range	Children
Academic	Man	Married	Early 50s	4
Company employee	Woman	Separated	Mid 40s	2
Consultant	Woman	Single	Early 50s	None
Entrepreneur	Man	Married	Mid 40s	2
Farmer	Man	Married	Early 50s	4

Semi-structured interviews and online surveys – the only qualitative data collection techniques listed by Casper, et al. (2007) in their meta-analysis of the research methods used in work-family research – were chosen as data collection techniques, following a review that found that they are commonly used and often in combination, for example, by Drew and Murtagh (2005).

To explore the construct from the perspective of the participant, the research first engaged the participants in interviews that asked open questions which, in so far as possible, did not impose a frame on the phenomenon. The online survey then followed up with a series of questions, drawing on labels for defining factors developed following a review of meta-analyses on work-life balance by Kalliath and Brough (2008), Allen et al. (2013), and Casper et al. (2018). Example interview and survey questions are given in the appendix.

Key factors emerged from analysis of data with triangulation between the methods showing significant alignment.

7.2.2   Study B

This study was conducted in the UK and focused on executives in medium-sized businesses. Three of the participants were women and one was a man. All were goal-oriented individuals who are subject to high levels of professional pressure and stress, working in different industries. This study was a phenomenological one that explored the phenomenon of work-life balance, using both in-depth, semi-structured, *elite interviews* (Kakabadse and Louchart 2011) and participant observation to collect data (Cassell and Symon 2004).

During the interviews, all questions started open-ended and were followed-up on with factual questions – rather than "opinion" ones – and other probing questions to gain a deeper understanding (Arksey and Knight 1999). Examples of questions asked are given in the appendix. During the interview, the interviewer also observed the respondents, focusing on the behavioural patterns of what was being said and done (Bryman and Bell 2015). There was an alignment between the key factors that emerged from the analysis of data collected by both methods and the findings were combined.

7.2.3   Study C

This study, like Study B, was conducted in the UK but focused on senior Human Resource (HR) professionals – both male and female at similar levels in the organizational hierarchy – in a single UK-based multinational company. All participants had their own view on their individual work-life balance as well as a deep understanding of the company's work-life balance policy, having both familiarity with its use in practice and the ability to influence it.

Semi-structured interviews and a 20-statement exercise, where participants are asked to write down 20 statements in response to the question "what are the key factors determining work-life balance," were used for data collection. The 20-statements allowed the participants to voice their thoughts on what they believed the key factors determining work-life balance to be, without being influenced by questions sourced from literature (example questions are given in the appendix). Its use was followed by semi-structured interviews which probed their thinking based on themes identified in the literature including, for example, Sturges and Guest (2004), Haar et al. (2014), and Anderson et al. (2017).

The analysis was aided by the use of the Symbolic Interactionism Framework (Snow 2001) and of Wilber's quadrants model (Wilber 2007). Key factors emerged from the data with triangulation between the methods showing significant alignment.

7.3 KEY FACTORS OF WORK-LIFE BALANCE

As human beings, we are constantly pulled in multiple directions – needing to balance the demands of work life, home life, and social life (Roche and Haar 2013). This multifaceted struggle has been explored by many researchers, and numerous different factors have been identified and examined. Building on previous research, the authors' studies find commonalities and differences from which a set interrelated key factors emerge.

7.3.1 FINDINGS OF THE INDIVIDUAL STUDIES

The two first studies explored work-life balance from an individual perspective, while the third supplemented this with an organizational dimension. The findings from the latter study could be considered to be a more restricted investigation into work-life balance than the others. However, even with this constraint, a number of common factors emerged from all three studies and the interplay between factors such as well-being, occupational stress, and flexibility and control was evident.

Study A found that participants did not consider work-life balance to be only about time but also about "cognitive demands," stress, and the ability to be "present in the now" in whichever domain they are in – their overall physical, mental, and emotional well-being. It also found that a sense of control and flexibility over where, when, and what one works on was important as was a person's individual circumstances such as their financial security, personal and parental status, the nature of their job and, where applicable, organizational culture (related to this was the finding that balance is not fixed or static). In all, six key factors emerged which were labelled; well-being, flexibility and control, time-use, occupational stress, individual circumstances and, most significantly of all, the individual's belief system (their values, attitudes, and beliefs) as this heavily influences their perception of what balance is and their approach to maintaining it.

Study B found health to be a recurring theme, brought up by all participants, as was the importance of having an active strategy to achieve a desirable work-life balance. These emerged as key factors labelled "health" and "self-management," respectively. The study also found that relationships and relationship dynamics (including how much flexibility one is given by the other party) influences work-life balance and how people feel about themselves. The impact of fulfilment also emerged as key since, for example, individual accomplishments (which can be achievements of different magnitudes that allow the individual to feel better about themselves) can trigger positive emotions and give rise to a sense of fulfilment. The findings also indicate that the notion of feelings and how we feel as individuals influence several aspects of our lives, thus impacting our work-life balance. How an individual's main priorities influence their behaviour and self-awareness also emerged as important factors. In all, seven key factors emerged which were labelled: health (physiological and psychological), self-management, relationships, fulfilment, feelings, behaviour, and self-awareness.

Study C found managerial support to be a critical factor, both from the perspective of ensuring an employee's workload is realistic and that of ensuring that employees feel valued and not judged for their flexible working pattern. It also found that this support, in turn, affects people's well-being and capacity and motivation for work. The trust between manager and subordinate emerged as important as it enables employees to be measured by their output instead of by their time in the office and encourages the use of the company's flexible working policy. The study also found that the collective factors of culture and their influence on organizational policies such as flexible working hours (which the study participants felt was a key enabler of balance) are important and moderated by attitudes towards gender equality and parental leave as inter-related factors. As part of a collection of individual factors, an individual's motivations and belief system affected their perception of their capacity for work. There was also a relationship between their sense of well-being and prioritization of work, which was a matter of individual choice.

In all, four key factors emerged, which were labelled: managerial support, organization, societal factors, and individual factors, one of which is an individual's belief system. The study also found that work-life balance was subject to change with, for example, changing family commitments.

7.3.2 CONSOLIDATION OF FINDINGS

Considering the findings of the research, it is apparent that work-life balance means different things to different people. One research participant noted that balance "is subjective to each person," while another said that "it's different for everyone" and "a personal thing" and a third referred to it as "an internal experience." However, after analysis and triangulation of the studies, which involved a lot of discussion and reflection on coding and themes, key factors that influence work-life balance emerged. They are the overlapping themes of well-being, individual circumstances, self-management, relationships, and an individual's belief system. These are illustrated in Figure 7.1 and are discussed, along with several subthemes, in the following sections.

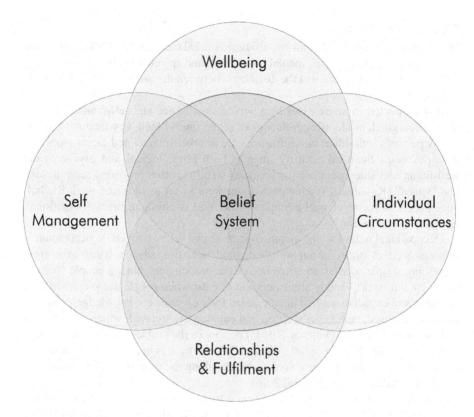

FIGURE 7.1   Key factors of work-life balance.

7.3.2.1 Belief System

We see that increasing workloads, long working hours, and unexpected events create insecurities (Brynien and Igoe 2006), which can throw people off balance and affect their quality of life. Whether a person is in or out of balance impacts their surrounding environment and so influences the people around them. However, since we are all different as individuals, we all perceive the world around us in different ways and have our own personal sense of reality. The stories we tell ourselves to make sense of the world around us can be thought of as a *belief system*.

A person's belief system – their values, attitudes, and beliefs – is the lens through which they see and interpret the world. As such, it heavily influences their perception of what balance is and their approach to maintaining it. It extends beyond each individual's constructed reality – as, in effect, one's belief system is intrinsic to how people construct their social reality – so is perhaps the most significant and generalizable of all the factors and two of the three studies identified it as being one of the most important key factors (this is illustrated in Figure 7.1, where the central circle feeds into the other key factors).

As belief systems vary, defining work-life balance is a difficult and complex task with many influencing and interrelated variables. However, commonalities and patterns, which we discuss below, emerged during analysis of the three studies.

7.3.2.2   Well-Being

Well-being, as it emerged from the studies, is a balanced state of life satisfaction which encompasses physical, mental, emotional, and spiritual health. It is so deeply linked to work-life balance that the distinction between the two was blurred for many participants.

It is supported in many cases by a person's resilience and self-management so the bidirectional, moderating influence of other, intertwined, key factors is apparent. A person's individual circumstances are another interrelated factor since, for example, one's financial security impacts both physiological and psychological well-being and when people have financial stability, other problems seem to take less "space" (Kalish 2014). One participant went as far as to frame work-life balance purely in terms of well-being "whether that is physical, mental, emotional, or material well-being."

Occupational stress – the psychological stress related to one's occupation – imposes a direct strain on our well-being and work-life balance. It can arise from something simple such as, to quote one of the participants, having people "breathing down our necks" and is almost certain to exist when people are presented with demands and expectations that do not match their capabilities, knowledge, or capacity. If not defused, people get worried and emotionally stressed, which can result in depression and burnout (Valcour 2016). The core of the problem, often found within ourselves, is rooted in our belief systems. This is why it is important to engage with oneself, to reflect and do some soul searching along the way as part of a regime of regular mental clean-ups (Paliwal 2016).

Related to stress is the notion of fun, which appears to affect both work-life balance and work performance. Having fun at work can increase engagement and productivity and is directly related to organizational citizenship behaviour, and indirectly to creativity and task performance (Fluegge-Woolf 2014). As is made clear by Csikszentmihalyi (1990), there are implications here for organizations when it comes to motivation and role design. If people do things that they are passionate about, it is intrinsically motivating and they enjoy what they are doing which may lead to better work performance (Csikszentmihalyi 1990) and a better handling of the pressure that follows (Roche and Haar 2013). Consequently, it enables a better work-life balance.

One study used Wilber's (2000, 2007) quadrant model to explore the interplay between individual, organizational, and societal factors. It found that, even though individual definitions of well-being vary, there was a clear relationship between individual-inner motivations of individuals feeling valued by their manager, their mental state, and a sense of well-being. These individual factors, in turn, lead to prioritizations of activities and, ultimately, to the exterior-individual choices made between different competing demands in different areas of our lives. Thus, in an attempt to keep the workforce healthy, many organizations implement "wellness programmes" despite arguments that they merely contribute to the problem by increasing pressure since such initiatives can be seen as an invasion of private life that disrupts work-life balance. Some also feel that they create a new platform for performance and a resulting *fear of not measuring up* (Berinato 2016).

7.3.2.3 Self-Management

Not surprisingly, time emerged as an important component of what people consider the influences on work-life balance to be. However, as noted by a participant
in one study, work-life balance "is about mental hours" and "if a person can't sleep
because of worrying about work duties, that is a problem, no matter how reasonable
the hours." It is not merely an issue of time management but of self-management,
which entails "(a) self-assessment, (b) goal setting, (c) time management, and (d) self-
regulation" (Gerhardt 2007, 11).

Tulgan (2017, 48) considers the essence of self-management to be "time management, interpersonal communication, organizational skills, basic problem solving."
Skills which we would like to see in organizations, team members, colleagues and, of
course, ourselves. Mastering them helps us improve work-life balance as, from time
to time, most of us experience problems concentrating and in staying focused as we
are subject to multiple distractions. Fortunately, they are skills that can be developed
with proper training and practice.

Self-discipline is another contributing factor to self-management. It involves managing yourself – including your emotions – and adapting your behaviour to achieve your
desired goals (Duckworth 2009). Thinking of self-discipline in this manner enables the
prediction of a variety of positive outcomes (Duckworth 2009) and contributes to work-
life balance. For example, one participant noted that she has a "personal tendency to
work too much," which she could counter with greater self-discipline.

Self-management, as was the case with well-being, is intertwined with other factors. For example, our sense of autonomy – of control over our own lives – impacts
our capacity for self-management. This is driven by the degree of flexibility allowed
by each individual's circumstances and how much control we have over where, when,
and what we work on as well as by the degree of control we have over our work and
non-work lives, both in terms of autonomy and the ability to plan. For example, one
participant remarked that he can set his own hours while another commented that
work-life balance means that she can maintain a sense of control over her life. In the
words of a third participant, "it's very difficult to strike the correct balance for your
own situation without having a degree of control" or, as it was put by another, to
achieve balance, "you need to plan it."

Having life goals and a sense of purpose helps people maintain work-life balance.
This has organizational implications, especially when it comes to motivation, where
intrinsic drivers are more favoured than extrinsic ones which are associated with burn-
outs. People who do things they are passionate about have a lower risk of burnout (Roche
and Haar 2013). However, even though one may be pursuing something one is passionate about, it is important to engage in continuous learning and personal development, as
stagnation also seems to be associated with burnout. For an organization, this highlights
the importance of empowering, inspiring, and motivating the workforce.

While having a purpose can help us maintain work-life balance, we should not
think that "purpose is a single thing" or that "purpose is stable over time" (Coleman
2017, 1). One should not seek to find a single purpose but to provide purpose to
everything one does while realizing that one's purpose may change over time
(Coleman 2017).

7.3.2.4   Individual Circumstances

Everyone experiences things differently and we all are constantly juggling differ-ent situations, relations, and, emotions depending on our individual circumstances. These circumstances range from financial security, personal and parental status, the nature of one's job, luck, and – where applicable – the organizational culture in which one works. The research found them to have a significant influence on how partici-pants viewed balance and so they emerged as major defining factors. For example, a participant from one of the studies believes that "the view of your employer is extremely important" while another noted that on certain days, she had a "com-plete clash of personal responsibilities" which conflict with writing requirements that "leak across seven days a week."

Since circumstances change, the related finding that balance is not fixed or static also came to the fore. For example, some participants with children indicated that their definition of balance changed once they had others depending on them. One of them noted that over the years, as he got wiser, what he wanted from life changed, while another commented that her demands for balance were now healthier than they used to be.

It further emerged that balance does not require things to be equally weighted but rather, to quote one participant, as "suits the individual circumstances which natu-rally will be different for everyone."

For those who work in organizations, organizational culture is a fundamental part of their individual circumstances. A participant in one study was quick to point out that the "culture in an organization matters" and that he currently had very little chal-lenge maintaining work-life balance. This contrasted starkly with a comment from another participant who, despite been given great flexibility by her employer, finds maintaining balance an "ongoing battle."

From an organizational perspective, there is no doubt that people are affected by their environment and it influences motivation, occupational stress, workforce capac-ity, and work-life balance. Bad conditions at work create environmental stress which amplify psychological reactions and potentially lead to irritation, headaches, and lack of focus (Lamb and Kwok 2016), damaging people's quality of life.

One approach to help manage individual circumstances is to nurture what can be called a "culture of health." A culture of health has been defined as one in which "individuals and social entities, such as households and businesses, can make healthy life choices within an environment that promotes options for health and well-being for everyone, and where the healthy choice becomes the valued and easy choice" (Koh 2018). It encourages individuals and organizations to maximize good health and well-being (Koh 2018) and so, facilitates organizations staying vibrant and com-petitive while enabling a healthy work-life balance.

Given the increasing challenges people face in maintaining a healthy work-life bal-ance, organizations that wish to succeed need to consider incorporating health as a com-ponent of their cultures. This culture, through elements such as manager support, policy, and working practices will have significant influence on work-life balance.

The research found that interior-collective factors such as trust, gender egali-tarianism, societal changes, and organizational culture to be important. However, the role of exterior-collective organizational policies, while identified as an enabler of work-life-balance, was consistently not regarded as being effective in isolation. Organizational policy is intermediated by both the interior-collective working culture and the exterior-collective working practice of an organization and gender egalitarianism. A manager's role in enabling work-life balance is important and trust is vital as it enables employees' performance to be measured by their output rather than by their input of time. This, in turn, encourages the use of flexible working policies by employees.

7.3.2.5 Relationships and Fulfilment

Relationships emerged as an important factor determining work-life balance not only, as noted above, because participants with children indicated that their definition of balance changed once they had others depending on them but also since having good and healthy relationships can bring support and serenity to a person's life (Essaid 2017). For example, being in a good and loving relationship is healthy. It helps people to stay balanced, become more effective at work, and helps them to "switch off" when out of the office or at home. Hence, having love in life facilitates a better work-life balance (Essaid 2017).

While participants were quick to disagree that relationships placed a burden on them (all the participants of Study A disagreed without hesitation), relationships do have a flipside as they introduce relational commitments with ongoing demands. Relationships can suffer due to occupational demands, stress, and workload, resulting in people feeling that they are not measuring up or coping with the increasing pressures of work life, home life, and social life (Roche and Haar 2013). It is also possible for relationships to negatively impact work-life balance as was clearly evident in a comment from a participant about how work suffered while he was getting separated from his wife.

Individual circumstances can positively and negatively affect how people feel and their emotional reactions towards people in their circle. Feelings and emotional patterns are closely connected to love for others or ourselves, which support better work-life balance. In contrast, occupational stress often triggers negative feelings which negatively influences individual well-being (Brynien and Igoe 2006). These feelings drive behaviour and how we manage ourselves, which is closely related to our level of self-awareness and how well we connect with other people.

Organizations need to be aware that as people progress in life, they experience increasing conflict between home and work life due to increasing pressure and demands from partners and family commitments (Sturges and Guest 2004). Human Resource professionals should plan accordingly for the changing needs of the younger members of their workforce, adapting organizational culture, policies, and working practice as required to accommodate the changing expectation of younger generations and supporting the call for work-life balance (Anderson, Baur, et al. 2017; Festing and Schäfer 2014).

7.4 CULTIVATING WORK-LIFE BALANCE

From the three studies – supported by underpinning literature from theorists, researchers, and business professionals – we see how belief systems, well-being, self-management, individual circumstances, relationships, and fulfilment are the key factors that define and influence work-life balance. However, how can work-life balance be cultivated?

As a person's understanding of balance is driven by how they see reality, the first implication of the findings is that work-life balance can be cultivated by nurturing a healthy belief system, built on our values, one which serves to guide not only how we interpret things but also our approach to decision making.

Self-management is closely linked to this and, here, self-reflection to improve self-awareness and self-discipline are actionable steps that can help cultivate work-life balance. Other simple steps that we can take include such things as focusing on a few selected goals while avoiding distractions.

It is also possible to cultivate work-life balance by working on well-being by, for example, ensuring that one has adequate physical activity, nutrition, and sleep in order to avoid the detrimental physiological and psychological impacts that would otherwise result.

Not all a person's individual circumstances will be something that can be influenced directly. However, by working on our belief system and self-management and by looking after our well-being and our professional and private relationships, we can equip ourselves to handle or at least partly mitigate circumstances which might otherwise adversely impact our balance.

The intertwined nature of the factors is such that truly nurturing work-life balance calls for a holistic approach to managing work-life balance. An approach which includes an individual's entire ecosystem ranging from internal factors, such as their belief system, to external factors, such as their relationships, and the organizational culture at work. Organizations have a role to play here. They need to step up and perform it well not only for the benefit of their people but for that of all their stakeholders.

From an organizational perspective, some of the steps that can be taken to cultivate work-life balance include ensuring that suitable policies are in place (for example, relating to flexible-working hours, remote working, parental leave) and striving to create a culture in which there is empathy and trust between management and staff as these positively influences work-life balance (Seppala 2016). In addition, steps should be taken to ensure that performance is understood in terms of outputs, not inputs such as time in the office.

The payback from these measures will not only benefit the organization's people but also the organization itself as even positivity and subjective well-being alone have been shown to make a positive impact, which can result in "1. better health, 2. lower absenteeism, 3. greater self-regulation, 4. stronger motivation, 5. enhanced creativity, 6. positive relationships and 7. lower turnover" (Tenney, Poole and Diener 2016, 27).

7.5   CONCLUSIONS

Although work-life balance was clearly found to mean different things to different people, the research found that there are a number of key factors, which we have discussed, that contribute to it. The nature of the research is such that care needs to be taken before generalizing the findings to other contexts. However, in order to do so, the studies can be built on by research in other contexts. In the meantime, it is possible for each of us, in our own context, to reflect on the research findings presented in this chapter and consider how well they help inform and explain our own understanding of work-life balance.

APPENDIX

STUDY A SAMPLE INTERVIEW QUESTIONS

1. What does work-life balance mean to you?
2. What are your personal challenges with maintaining a work-life balance?
3. Can you tell me about times you felt you work-life balance was lost or threatened? *Probe*: Why do you feel this was?
4. What did you do to regain a feeling of work-life balance?
5. Is there anything else that you think is important about work-life balance?

STUDY A SAMPLE SURVEY QUESTIONS

1. Please weight (between 0% and 100%) the following 3 potential components of work-life balance as they apply to you: Time balance, Involvement balance, Satisfaction balance. What is the reason for your weighting?
2. Consider the following possible definition of work-life balance and explain why you agree or disagree with them:
   • Balance involves equality of time or satisfaction between your multiple life roles
   • Balance involves satisfaction and good functioning at work and at home with minimum role conflict
   • Balance requires work and life to be equally weighted

STUDY B SAMPLE INTERVIEW QUESTIONS

1. How would you define work/life balance?
2. How would you describe your own work/life balance?
3. How does your social life play a part in this?

STUDY C SAMPLE INTERVIEW QUESTIONS

1. What are the key factors in defining work-life balance, make twenty statement that define your perspective?
2. What are the most significant factors in defining work-life balance for you as an individual?
3. What level of organizational support is there for work-life balance in your organization? *Probe*: What do you think that means for your company? *Probe*: How do you feel about that?

4. In what way do policies providing organizational support for work-life balance affect an organization in today's context? *Probe*: Why do you think that? *Probe*: What do you think that means for your company?

REFERENCES

Allen, T. D., R. C. Johnson, K. M. Kiburz, and K. M. Shockley. 2013. "Work-family conflict and flexible work arrangements: Deconstructing flexibility." *Personnel Psychology* 66 (2): 345–376.

Anderson, H. J., J. E. Baur, J. A. Griffith, and M. Ronald Buckley. 2017. "What works for you may not work for (Gen) Me: Limitations of present leadership theories for the new generation." *The Leadership Quarterly* 28 (1): 245–260.

Arksey, H., and P. Knight. 1999. *Interviewing for Social Scientists*. 1st ed. London: Sage Publications Ltd.

Berinato, S. 2016. "Corporate wellness programs make us unwell: An interview with André Spicer." *Harvard Business Review* 93 (6): 29–28.

Blaikie, N. 2009. *Designing Social Research*. Cambridge: Polity.

Bryman, A., and E. Bell. 2015. *Business Research Methods*. 4th ed. Oxford: Oxford University Press.

Brynien, K., and A. Igoe. 2006. *Occupational Stress Factsheet*. New York: New York State Public Employees Federation.

Carlson, D. S, K. Michele Kacmar, and L. J. Williams. 2000. "Construction and initial validation of a multidimensional measure of work–family conflict." *Journal of Vocational Behavior* 56 (2): 249–276.

Casper, W. J., L. T. Eby, C. Bordeaux, A. Lockwood, and D. Lambert. 2007. "A review of research methods in IO/OB work-family research." *Journal of Applied Psychology* 92 (1): 28.

Casper, W. J, H. Vaziri, J. H. Wayne, S. DeHauw, and J. Greenhaus. 2018. "The jingle-jangle of work–nonwork balance: A comprehensive and meta-analytic review of its meaning and measurement." *Journal of Applied Psychology* 103 (2): 182.

Cassell, C., and G. Symon. 2004. *Essential Guide to Qualitative Methods in Organizational Research*. London: Sage.

Chang, A., P. McDonald, and P. Burton. 2010. "Methodological choices in work-life balance research 1987 to 2006: A critical review." *The International Journal of Human Resource Management* 21 (13): 2381–2413.

Coleman, J. 2017. *You Don't Find Your Purpose - You Build It*. Harvard Business Review Digital Articles.

Crotty, M. 1998. *The Foundations of Social Research: Meaning and Perspective in the Research Process*. St Leonards, NSW: Sage.

Csikszentmihalyi, M. 1990. *Flow: The Psychology of Optimal performance*. Vol. 40. New York: *Cambridge University Press*.

Drew, E., and E. M. Murtagh. 2005. "Work/life balance: Senior management champions or laggards?" *Women in Management Review* 20 (4): 262–278.

Duckworth, L. A. 2009. "Backtalk: Self-discipline is empowering." *The Phi Delta Kappan* (The Phi Delta Kappan) 90 (7): 536–536.

Easterby-Smith, M., R. Thorpe, and P. R. Jackson. 2015. *Management and Business Research*. London: Sage.

Emslie, C., and K. Hunt. 2009. "'Live to Work' or 'Work to Live'? A qualitative study of gender and work–life balance among men and women in mid-life." *Gender, Work & Organization* 16 (1): 151–172.

Essaid, R. 2017. "How falling in love made me a better CEO" *Fortune* 1–1.

Festing, M., and L. Schäfer. 2014. "Generational challenges to talent management: A framework for talent retention based on the psychological-contract perspective." *Journal of World Business* 49 (2): 262–271.

Fluegge-Woolf, E. 2014. "Play hard, work hard: Fun at work and job performance." *Management Research Review* 37 (8): 682–705.

Gerhardt, M. 2007. "Teaching self-management: The design and implementation of self-management tutorials." *Journal of Education for Business* 83 (1): 11–17.

Grady, G., and A. M. McCarthy 2008. "Work-life integration: Experiences of mid-career professional working mothers." *Journal of Managerial Psychology* 23 (5): 599–622.

Greenhaus, J. H., and N. J. Beutell. 1985. "Sources of conflict between work and family roles." *Academy of Management Review* 10 (1): 76–88.

Gioia, D. A., K. G. Corley, and A. L. Hamilton. 2013. "Seeking qualitative rigor in inductive research: Notes on the Gioia methodology." *Organizational Research Methods* 16 (1): 15–31.

Haar, J. M., M. Russo, A. Suñe, and A. Ollier-Malaterre. 2014. "Outcomes of work–life balance on job satisfaction, life satisfaction and mental health: A study across seven cultures." *Journal of Vocational Behavior* 85 (3): 361–373.

Kakabadse, N. and E. Louchart. 2011. "Delicate empirisicm: An action learning approach to elite interviewing." In *Global Elites: The Opaque Nature of Transnational Policy Determination*, eds. A. Kakabadse and N. Kakabadse, 286–307. London: Springer.

Kalish, B. M. 2014. "New wellness programs focus on finances, not just fitness." *Employee Benefit News* 28 (2): 8–9.

Kalliath, T., and P. Brough. 2008. "Work–life balance: A review of the meaning of the balance construct." *Journal of Management & Organization* 14 (3): 323–327.

Koh, H. 2018. In Improving Your Business through a Culture of Health, produced by HarvardX, Edx course video, 01:29, https://www.edx.org/course/improving-your-business-through-a-culture-of-healt.

Lamb, S., and K. C. S. Kwok. 2016. "A longitudinal investigation of work environment stressors on the performance and wellbeing of office workers." *Applied Ergonomics* 52: 104–111.

Lewis, S., and T. Alexandra Beauregard. 2018. "The meanings of work-life balance: A cultural perspective." In *The Cambridge Handbook of the Global Work-Family Interface*, eds. W. Shen R. Johnson and K. M. Shockley, 720–732. Cambridge: Cambridge University Press.

Paliwal, D. 2016. "Harman CEO: 5 ways to avoid burnout." *Fortune* 759–759.

Roche, M., and J. M. Haar. 2013. "Leaders life aspirations and job burnout: A self-determination theory approach." *Leadership & Organization Development Journal* (Emerald Group Publishing Limited) 34 (6): 515–531.

Saunders, M. N. K., and K. Townsend. 2016. "Reporting and justifying the number of interview participants in organization and workplace research." *British Journal of Management* 27 (4): 836–852.

Seppala, E. 2016. "Good bosses create more wellness than wellness plans do." *Harvard Business Review Digital Articles* 2–4. Accessed: 7 June 2018, https://hbr.org/2016/04/good-bosses-create-more-wellness-than-wellness-plans-do.

Snow, D. A. 2001. "Extending and broadening Blumer's conceptualization of symbolic interactionism." *Symbolic Interaction* 24 (3): 367–377.

Sturges, J., and Guest, D. 2004. "Working to live or living to work? Work/life balance early in the career". *Human Resource Management Journal*, 14 (4): 5–20.

Tenney, E. R., J. M. Poole, and E. Diener. 2016. "Does positivity enhance work performance? Why, when, and what we don't know." *Research in Organizational Behavior* 36: 27–46.

Thilagavathy, S., and S. N. Geetha. 2020. "A morphological analyses of the literature on employee work-life balance." *Current Psychology*: 1–26. Doi: 10.1007/s12144-020-00968-x

Tulgan, B. 2017. "Teaching the fundamentals of self-management." *Talent Development*, September: 48–52. Accessed: 15 February 2018, https://www.td.org/magazines/td-magazine/teaching-the-fundamentals-of-self-management.

Valcour, M. 2016. "Beating burnout." *Harvard Business Review* 94 (11): 98–101.

Wilber, K. 2000. *Integral Psychology: Consciousness, Spirit, Psychology, Therapy.* Boston: Shambhala Publications.

Wilber, K. 2007. *The Integral Vision: A Very Short Introduction to the Revolutionary Integral Approach to Life, God, the Universe, and Everything.* Boston: Shambhala Publications.

8 Smart Cities and Their Role in Enhancing Quality of Life

Uday Chatterjee
Bhatter College

Gouri Sankar Bhunia
TPF Gentisa Euroestudios

Dinabandhu Mahata
Central University of Tamil Nadu

Uttara Singh
CMP Degree College (University of Allahabad)

CONTENTS

8.1   INTRODUCTION

Since the 1990s, the word "Smart City" has been in use and different meanings illustrate numerous facets of the term. Such concepts require the use of ICT (Information and Communication Technology) (i.e., bricks) to include people, to provide public infrastructure, and to improve urban structures through "Smart Cities". Today, half of the population lives in cities, and by 2050, as predicted by the United Nations, the world's population will grow from 7 to 9 billion; simultaneously the urban population will double in size. The direction of rapid urban population growth is not only fascinating but calls for a compelling imperative for economic development and better livelihood.

The rapid growth of cities and their disproportionate use of physical and social resources is unsustainable, just as traditional systems, in which a city's supply resources are dependent. Cities today face challenges such as unimpeded and unplanned growth, scarcity of physical and social services, environmental demands and regulatory restrictions, reduced tax base and budgets, and increased cost of living (Kumar and Jailia, 2018). Thus, there is a need to identify innovative and creative strategies for addressing urban life challenges and issues ranging from congestion, overpopulation, metropolitan growth, high unemployment, allocation of land, environmental security, and increasing crime rate. Inadequate residential spaces especially in low-income communities, coupled with dearth of infrastructure and a variety of essential services, viz potable water supply, sanitation, education, and health, are long-standing urban challenges. Since a sizeable population will be based in and around these urban centres, the quality of our future towns will determine the future of our world.

The predicted exponential growth in urban population coupled with the rising aspirations of these populace necessitates the implementation of contemporary and futuristic technologies with minimum human intervention. Human intervention has to be minimized because of the inherent biases of humans, vulnerability to socio-political milieu and other intangible factors. Since these urban centres will be the drivers of future growth and will form the bedrock of a nation's overall growth, it, therefore, is absolutely essential that the functioning of such urban centres has to be automated to an extent which leads to optimum and efficient management of all hallmarks of a city life, viz infrastructure, resources, health care, traffic management, disaster management, planned expansion, education, sustainability and overall quality of life. Therefore, to ensure that these urban centres don't just become "population centres" which are grossly mismanaged, ill planned and become counterproductive to the growth of the economy, the use of digital information technologies is an essential tool for urban planning and participatory decision making in the planning process (Yigitcanlar, 2020. A "Smart City" is designed to optimize the quality of life of residents through the use of technology and the integration of various key functions such as data management, smart transport, and security. The vision "Smart Cities" will be a futuristic urban centre, which is safe, secure, green, efficient and based on advanced integrated materials, sensors, and networked environment which will be tracking and networking interfaces of all the civic facilities like electricity, water or transport, etc. In fact, smart city implementations are made of a variety of apps and state-of-the-art technologies (ICT implementations). The concept of the Internet of

Things (IoT) is a critical component in shaping future cities, which will be implemented by various devices like sensors, gateways, communications infrastructure, and servers. The IoT is a global infrastructure for the information society, enabling advanced services by interconnecting (physical and virtual) components based on existing and evolving interoperable information and communication technologies. By connecting physical and virtual platforms which contain embedded technology to communicate and sense or interact with their internal states or the external environment, data can be extracted from physical objects and machines. Subsequently, these data can be aggregated, analysed, leveraged for intelligence, decisions, and applications, including autonomous actions by the connected devices themselves. Obviously, IoT will form the backbone of any futuristic and efficient smart city.

The use of technology from "Smart Cities" results in cost-efficiency, a robust and resilient infrastructure, and an enhanced urban experience. "Intelligent Cities" are the latest concept for the building of future cities. "Smart Cities" will play a key role in combining a sustainable future with the continuous growth of the economy and job creation in order to bring new identity and a unique value to the lifestyle of its inhabitants. A city's global competitive benchmarking, good governance, and the best possible provision of municipal services will be some of the characteristics of "Sustainable" intelligent cities in the new era. New developments around Smart City now include eco-towns, urban cities, sustainable communities, and last but not the least "aerotropolis".

Smart Cities have a very fair and clear end goal, which is to enhance the quality of life of its inhabitants and simultaneously improving their functional efficiency. This initiative can rapidly bring about cost-effectiveness to trade, transport, services, education, and housing provide better living environment and enhance security setup in the cities. A city's infrastructure consists of housing, drainage, sewage, water supply, power generation and distribution, transportation, waste management, and connectivity among others (Joseph, 2014). In order to achieve an improved performance, an intelligent city infrastructure differs from the conventional urban infrastructure in its capability to respond intelligently to environmental variations, including user requirements and other infrastructural changes. Smart city infrastructure is the foundation for all six key identifiers of intelligent cities: intelligent mobility, intelligent economy, intelligent living, intelligent management, intelligent people, and smart environment. Yet the intelligent elements of the network are extremely context-sensitive and their design depends on the stage of growth and the unique planning problems in the cities. In fact, these intelligent network systems will provide the basis for new technologies which will provide an ecosystem conducive for productivity and optimized resource management. The loss of basic functionality at a living location indicates problems arising from rapid development, e.g., difficulties in waste management, scarcity of resources, air pollution, adverse effects on human health, traffic congestion, and insufficient and dilapidated infrastructure.

8.1.1 Globalization and Smart City

Barcelona, Copenhagen, Singapore, London, Seoul, and Helsinki are some of the cities which have been recognized for their implementation of efficient smart solutions, e.g., Barcelona is a global leader in the IoT and has successfully integrated smart street lighting solutions and intelligent energy usage. Likewise, Copenhagen transfers less than 2% of its waste to landfills; half is recycled and the majority is used as input to the thermal grid in its heating network. Similarly, Singapore's all government services are accessible online and have several citizen-focused mobile apps for transport, health care, and municipal services which were developed by an in-house digital government team. In the last few decades, cities have experienced dramatic changes due to the pressure from a vast population concentration, dominance of vehicular movement, and inappropriate urban planning approaches (Saeidi and Oktay, 2012).

By 2050, the world's population is expected to grow to more than nine billion, with 80% of it residing in urban areas. At present, over half of the world's seven billion inhabitants live in the cities. It is projected that the 600 largest cities in the world will account for 60% of world's GDP by 2025. It is anticipated that the 30 largest cities alone will lead to 20 % of global GDP growth between 2010 and 2020. In the meantime, cities account for only 2% of the land mass and consume 80% of the world's natural resources. The key element of a "Smart City" is its strategic application of new and high-level ICT solutions to connect the citizens and technologies on a shared platform, whereas traditional cities suffer from exponential growth and disproportionate consumption of physical and social resources which makes them unsustainable in future. Although urbanization results in rise in global carbon emissions, focused local efforts and smart solutions can stem this problem to a large extent. In fact, to tackle climate change, our cities are our best hope of combating chronic twin problems of "Global Warming" and the new era of "Green Economy" in food security. After the 2008 crisis, the intelligent development of harbour areas and cities was based on the following specific principles: the principle of synergy, the principle of creativity, and the principle of circulation (between different stakeholders/ systems, in particular, socio-cultural and economic systems) (Fusco Girard, 2013). The Historic Urban Landscape (HUL) reinforces the fact that the transition to an intelligent growth paradigm is based not just on technical advances, but on unique local cultural capital.

8.1.2 Sustainable Development and Smart City

Sustainable towns have become a widely desirable target for sustainable economic planning without harming the climate. A deeper understanding of the "Smart City" concept is needed to ensure healthy living conditions in this fast-paced urban population growth worldwide (Phadtare and Jadhav, 2017). Many cities around the world continue to find better ways of coping with these problems. The label "Smart City" describes these cities aptly which are an icon of sustainable and habitable ecosystem. As a way to organize and identify different forms of knowledge stored or circulated within the networks of "Smart Cities", IBM developed the Concept of Urban Knowledge. Using its multiple layers, with each layer reflecting a specific two-dimensional space, the terrain of the urban world, whether it is a single city or a metropolis, can be seen in the Urban Information System. People by connecting to the "Smart Cities" can contribute to the growth of their community; learn about goods, services, volunteer activities; and get in touch with other people with similar interests, all in their own time and in ways that facilitate connections. "Smart Cities" therefore mainly benefit from building a more connected and networked community.

In addition to developing and integrating new and intelligent technologies, innovative governance models too need to be incorporated in a "Smart City" system. Such policies must be less "Top-Down" like conventional governance policies and must rely more on horizontal governance, which facilitates mutual collaboration and networking among different stakeholders. These silos of knowledge must be opened up and integrated into each other for the future development of multifunctional Smart City solutions. Smart cities are therefore required to cooperate with various stakeholders, including producers, institutions of knowledge, citizens, and municipalities. Over the past decade, the advent of new technology has brought about the emergence of smart cities aiming to provide their stakeholders smart technology-based solutions that are effective and efficient (Khan, Woo, Nam, and Chathoth, 2017).

8.1.3 SMART CITY – AN INTELLIGENT CITY

Developing intelligent cities implies developing urban centres which are not only integrated, habitable, and sustainable but also incorporate all available technologies and resources, integrated into an intelligent and coordinated way. ICT opened up an entirely new dimension for urban development. The efficiency of ICT networks and their functionality are not the only concepts of a "Smart City". For example, Berry and Glaeser (2005) and Glaeser and Berry (2006) demonstrate that in cities where a high share of qualified labour is given, the highest economic growth rates are achieved. The relation between human capital and economic growth is developed, in particular, by Berry and Glaeser (2005), suggesting that creativity is driven by entrepreneurs who innovate in manufacturing and in goods that involve an increasingly professional labour force. For instance, e-government activities assisted by ICT allow its residents and external institutions, to access required e-services within the integrated business processes of an institution. Consequently, these services must typically be managed by the supplier organization, through the units of business and the officers who are responsible for business processes and ICT development. However, architectures: business architectures, architectures of information systems, technology, and processes used to produce them can reinforce and strengthen the area of service management. The technology and networks become constantly linked and the difficulty of communities is intensified as new approaches for developing, transmitting, and using data begin to evolve. Cities are increasingly using digital systems in order to communicate with citizens and stakeholders and use the data and information they provide for the planning and delivery of the service. In addition to the seamless connectivity, which enhances the proliferation of IoT, technological innovation can resolve the real pressures related to traffic jams, waste/ pollution management, and energy efficiency, leading to an increasingly healthy and efficient urban life cycle.

"Smart Cities" are always a work in progress, but cities all around the globe would get better acclaim if these initiatives have real business results and their unique needs are taken into account. Internet of Everything (IoE) community technology architectures needs seamless sensor integration in a dynamic communication environment.

A particular network, such as streetlight management, video surveillance, or environmental monitoring, is traditionally installed around a given application. While separate networks naturally separate domains, they are usually not optimized (costs, security, availability), which leads to silos of information. Today, the Internet, not to mention our cities, has entered almost every area of industry and culture. IoT and IoE are the main catalysts for smart urban growth. Cities benefit from the exponential advances in information and communications technology to connect people, processes, data, and things. In conjunction with analytical and cognitive intelligence improvements, cities develop tools and systems for better performance monitoring, detect patterns, predict trends, and visualize large quantities of information spatially. It paves the way to informed cities and the assists in taking better infrastructural decisions.

In order to make urban systems more economically and environmentally sustainable, "Smart Cities" use digital technology. Sensors built into buildings and utility networks can enable communities to integrate green resources such as solar resources or conserve electricity. Sensors such as smart and digital cameras can assist in decision making and improve urban management by feeding real-time information into integrated management systems and by improving data and analytical technologies. Hopefully, the policy document to be issued shortly by the Government of India will more clearly explain its vision of an intelligent Indian city. Singapore, a leading advocate for smart city development and planning, recently looked for clarification about the idea of a smart urban by the Indian government. There are examples in India and internationally, of the cities that which use low-cost and ground-up smart technologies to improve public services, map information systems, and promote public engagement. The technical parameters should be driven by issues they address and the contribution they make towards meeting the needs and aspirations of local citizens and potential investors. The concept instead of being taken as a "show window" should be taken seriously to produce jobs, environmental and social standards, effective governance, and improve public involvement in the functioning of the city. The number of total Indian urban population is estimated to reach 843 million by 2050. By the year 2050, India will need smarter ways of handling challenges, reduce spending, increasing productivity, and enhancing the quality of life to handle this huge urbanization. Nowadays, the term "Smart City" is very widely used in spatial planning literature or urban research (Giuffrè, Siniscalchi, and Tesoriere, 2012).

8.2   SMART CITY – AN INDIAN PERSPECTIVE

According to the 2011 Population Census data, in the past decade, urban India grew by 90 million people. Over the same period, 2,774 new villages were born with more than 90% of the new villages in the town category. As stated in the "Smart City" mission document of 2015 issued by the Ministry of Urban development, 40% of India's population is expected to be in urban areas and will account for 75% of India's GDP in 2030 (Ministry of Urban Development, Govt. of India, 2015). There is a need for comprehensive progress in the growth of physical, structural, social, and economic infrastructures. We are all necessary to boost life quality, draw people and investment, and launch a virtuous growth process (https://www.india.gov.in/-spotlight/

smart-cities-mission-step-towards-smart-india). Around one-third of the newly developed towns are settled near (within a 50 km distance) to old city / metropolitan areas in India. These suburbs which only occupy 1% of the land area of India, accounts for about 18% of the workforce of the country. The Smart City Mission is one of Prime Minister's (2015) most ambitious projects which was launched in 2015 and is now being implemented in 100 cities with central funding for infrastructure development and making smart choices. According to the Ministry of Urban Development, the definition of the "Smart City" is not universally accepted because it can mean something different for different states and different countries. In the Indian context, the goals of an intelligent city would be to provide people with a clean and sustainable environment with basic infrastructure and a decent standard of living by incorporating "Smart Solutions". The concept is to implement inclusive and sustainable development in compact areas and further implement the same in the other cities. India does not have many of the institutions necessary, such as a transparent land use converting system, a clear definition of property rights, a strong land, and property valuation system, and a strong judicial system to address public concerns in order to facilitate land markets, land transactions and changes in land usage. In the present chapter, we assessed the role of a "Smart City" in enhancing the quality of life with special reference to India.

8.2.1 SMART CITY IN SUBURBAN AREAS

Smart cities can assist in establishing residential areas by ensuring fair distribution and quick provision of essential public service entities such as a supermarket or a medical store in the areas that lack them. The domestic life hub of a city is its suburban areas and therefore must be the main focus of the administration with an aim to improve the quality of life in a city. Another way a smart, upcoming community can improve the life for its residents is to interweave cleanliness in the surroundings by adopting measures such as placing waste disposal bins at periodic intervals and enforcing a strict anti-littering code of conduct. It is a long and difficult road to set up an ideal residential area in a city and to succeed, sincere sustained efforts are necessary from both ends of the society, i.e., government and the citizens.

8.2.2 LAND VALUATION AND TRANSACTIONS

Rapid urbanization in recent years combined with a weak institutional framework governing land appraisal, transaction, and acquisition has created increasing pressure on urban centres which has resulted in a distortion to the pace and form of urban expansion. India has no such robust programme, e.g., to achieve land valuation independently and consistently. In the past, high stamp duties have led to malpractices which led to the under-valuation of the land and property. The rise or decline in housing prices, according to property developers, rely on the four main factors: cost of land, permit prices, cost of raw materials, labour cost, and taxes.

One of Indian city and urban local authorities' most important sources of revenue is property tax. Each Urban Local Bodies (ULB) faces problems with regards to accessing the correct detailed information of any property: location, burden, property tax details, and disputes. The tax information on the property is mostly never integrated with the information about the property, such as burdens, because the information was not shared with the registration department. As a result, to search for property information becomes a tedious and time-consuming process. The main purpose of the smart municipality was to integrate the geo-referenced data on the properties with property-tax data with ULBs that provide details of owners of the property, the addresses, details of the property, tax details with the image of the property and their location. Intelligent cities should be a catalyst for improving the quality of life and have a positive impact on the management and transaction of immovable assets/ properties. Smart city property mapping is very useful as it integrates geo-spatial information with the MIS database and can be used for a number of urban initiatives, including increased access to detailed revenues on property data and improved service levels. Faster service and improved citizen satisfaction of the ULB services will enable hassle-free transaction for the people, which will result in saving precious time and resources.

8.2.3 PUBLIC TRANSPORTATION IN SMART CITIES

The rapid growth of individual travel modes, low public transit driving, high public transport rates, and poor transport speeds have all challenged the urban mobility in the Indian cities. In addition, there is evidence of increasingly difficult access to local and regional markets for companies located in peripheries. In India, the freight rates between cities and outskirts amount to Rs. 5.2 per ton kilometre (Lall, 2013), twice the national average and five times the costs in the USA. The lifestyle scales can be specified by an efficient public transport network. The transport network and the traffic that moves to and fro between various parts of the city, is a major aspect of a city life. They must, of course, be operational, maintained regularly, and integrated into an efficient network in order to maximize their potential and facilitate a hassle-free experience to the commuters (Figure 8.1). A smart city should have a road network which links all remote parts of the city, allowing people with all economic backgrounds fast and easy access to

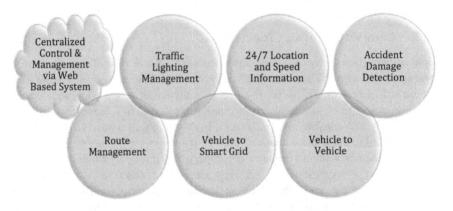

FIGURE 8.1   Components of an intelligent traffic system in a smart city.

facilities like hospitals, schools, and banks from their homes. For example, subway systems can coordinate their service and provide drivers with time estimates and alarms with the right array of sensors, digital signages, and mobile interoperability. Traffic flow monitoring and high traffic convergence points can contribute to the analysis and improvement of current road structure and organization. For instance, whether a beam-supported structure carries heavy traffic or large vehicles, resources may be selected to reinforce the supporting highway beams on a quarterly basis. The simple concept of collecting statistics on roads and traffic can make the citizens' travel more intelligent, practical, and safe. Intelligent Robust, strict, and efficient implementation of these technologies is essential in the effective functioning of a smart city.

8.3 DISASTER PROTECTION AND SMART CITY

A report by the United Nations states that almost 890 (60%) of people stay in cities which are vulnerable to at least one big natural hazard, including floods, droughts, cyclones, or earthquakes (https://ourworldindata.org/natural-disasters; United Nation, 2015). Eventually, any disaster can wipe off years of development and also cause death, injury, financial loss, and degradation of the environment and urban systems. Smart cities can use advanced ICT infrastructure and analysis capacity to enhance and coordinate information flows between multiple public institutions, such as transport agencies, emergency services, suppliers of energy, and citizens, as they face natural disasters (Japan Meteorological Agency, 2013). A community will meet the bulk of its residents shortly with the aid of mobile networks. Moreover, sensors can give advance warning of these disasters to all the citizens and help them cope up with the after effects in some situations (Figure 8.2). For example, IoT sensors are capable of gathering information, tracking conditions, and controlling systems. Insights are provided by large IoT data streams in real time (Shah et al., 2019). This leads to timely fact finding, efficient planning, and decision making, all of which play critical roles following a natural disaster (Figure 8.3). In order to prepare better for natural disasters, cities can leverage several data sources in real time. Drivers can plan ahead for the fastest and safest route while they get real time inputs of blocked roads. People can also calculate the best time to leave home for work. Similarly, flight information sharing in real time and advisories on emergency services can help mitigate the effects of any disaster. Connectivity offers various advantages in this field allowing for the extraction of situational information and the creation of robust and impactful disaster relief solutions (Chandra et al., 2016). In addition, interconnection empowers a community in more comprehensive planning, handling and rebuilding, post a natural disaster. Through the integration of IoT-based surveillance, private connectivity, and distributed infrastructure, governments and businesses can protect people and minimize economic losses due to any disaster. The insights that will be generated will also enable smart cities to improve the urban planning, development, and safety of all of their people. Over the last few decades, metropolitan areas worldwide have been engaged in a multitude of initiatives aimed at upgrading urban infrastructure and services (De Jong, Joss, Schraven, Zhan, and Weijnen, 2015).

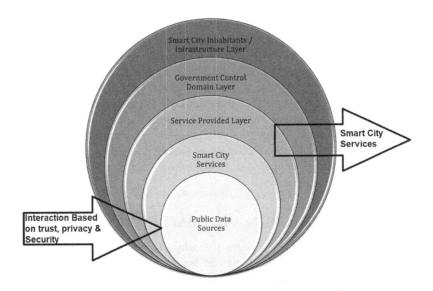

FIGURE 8.2   Smart city services in disaster management plan.

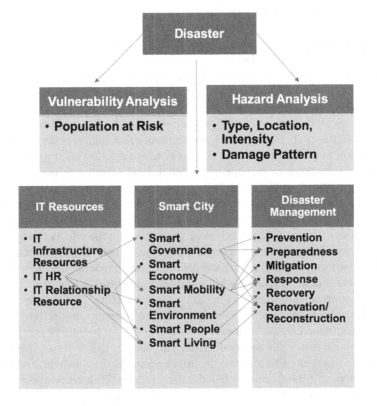

FIGURE 8.3   Smart city: utilization of IT resources to encounter natural disaster.

8.3.1 SECURITY AND SMART CITY

For cities around the world, public protection is a cause of great concern. The cities of the world are vandalized, civic resources are put under pressure, and checking the rising crime graph is becoming increasingly difficult for the police than ever before. Data storage and recovery issues can mean that retrieval of data from incidents can be slow for the investigating agencies and data loss may totally stall or hamper objective investigations (Li et al., 2020). In the recent times, the large and small districts are proposing a new city model, which represents a community development through smart city model of interconnected, sustainable, comfortable, and attractive, and secure (Lazaroiu and Roscia, 2012). Nevertheless, modern equipment has started to be make its way to the law enforcement personnel which will help them in dealing with such unlawful acts (Figure 8.4). In smart city solution, public safety begins by:

a. **Visual surveillance**: Reduces violence by helping police departments locate and arrest suspects which are likely to endanger public safety.
b. **Critical infrastructure monitoring**: Sensors embedded in critical infrastructures such as bridges and power plants log and transmit structural information to identify potential hazards thus protecting the citizens and the economy.

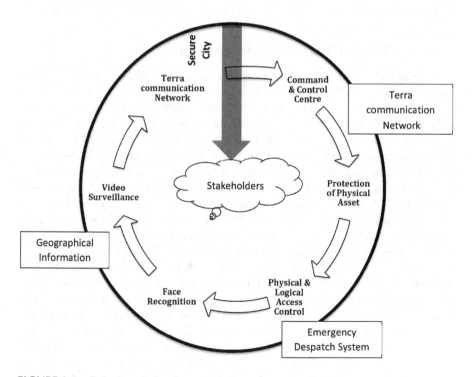

FIGURE 8.4   Safe city solution framework through smart technology.

c. **Police wearable**: Provides accountability and evidence to help with griev-
ances settlement and fosters constructive relationships which creates trust
between residents and police officers.
d. **Data security**: Ensures the safeguarding and accessibility of private and
other allied sensitive data collected to authorized personnel only.
e. **Environmental monitoring**: For better pollution control and minimize
the environmental impact, the sensor technology, and climate and pollution
data provided by citizens is used.
f. **Drone monitoring**: Vehicular flow, crowds, building, and emergency zones
are tracked by sensor equipped drones to assist first responders and record
prevailing conditions.

The Floor Space Index (FSI) regulations have placed the stringent cap on town-
density in Indian cities. The report argues that these rules lead to urban growth on the
outskirts of existing urban areas and encourage expansion. The FSI-induced growth
in Bangalore, for example, has been measured at 2%–4% of social security deficits
due to higher commuting costs. A Juniper report describes the use of a data-driven
approach by cities like New York or Chicago to build predictive models to help
police enforcement and emergency departments better target their resources (https://
newsroom.intel.com/wp-content/uploads/sites/11/2018/03/-smart-cities-whats-in-it-
for-citizens.pdf). According to a recent report by Bard College (2018), police are
researching emerging technologies such as drones, which have seen national security
authorities increase their usage by 82% in the last year alone (Gettinger, 2018).

8.4 IMPROVED UTILITIES AND SMART CITY

The research report shows that accessing services is based on wide core differences.
For instance, 93% of households in the core urban areas have access to drainage in
India's seven largest cities. At a distance of 5 kms from the core, this proportion drops
to 70%. Other essential services such as access to piped water in cities like Bangalore
are mainly concentrated in the urban core. It is easy to take for granted low costs of
utilities, clean air and water, and an efficient waste management system but citizens
are sure to notice a problem. With more pressure on these services than ever before,
IoT devices can offer highly efficient solutions. Intelligent water systems can optimize usage and minimize leakages while water sensors can identify and allow rapid
reaction to dangerous contaminants, including radioactive waste, heavy metals, and
even explosives. Rehabilitated structures, potholes, defective street lights, and faulty
traffic signals can cause be both a nuisance and even endanger human lives. In this
case, intelligent sensors in urban infrastructure such as bridges and streetlights might
automatically send maintenance requests to a centralized repair agency. Some scholars also pointed out the necessary ingredients for the composition of a smart city,
which is an as smart economy, smart mobility, smart environment, smart people,
smart living, and smart governance (De Jong et al., 2015; Lazaroiu and Roscia, 2012).

The Urban flow project has constructed an interactive series of kiosks across the
city that allow passers-by to check for streets and destinations (Gómez-Carmona
et al., 2019). The kiosks also chart the best roads, not only by distance but also by

walking time. If a user touches a location on the screen, it lights up and helps to separate the farmworkers (from the visual clutter that surrounds it.

8.5   HEALTH AND SMART CITY

Many initiatives have now been taken with the objective of fostering a broader view of health and well-being, which has led to a great deal of demand for smart wearable devices as fitness trackers or fitness groups and even health assessment applications on smartphones. Smart Health technology integrates and works with the data generated by the tools to be analysed for precise clinical treatment and suggests approaches to be taken by the physicians, experts, and healthcare professionals. In healthcare, IoT provides valuable insights by connecting information collected from intelligent devices and sensors. Urban landscaping sensors can be used for streamlining Emergency Response procedures and can detect events like car accidents and immediately alert the appropriate emergency service. With faster and more comprehensive access to specific patient data, Emergency Medical Services (EMS) teams and hospitals would be more successful in saving lives. It will also help to quickly assimilate data from the scans, track the patient's condition, and then relay it in real time to physicians and nurses, thereby enhancing the overall performance of the health care system. AI can also be used to perform activities such as field research, X-rays, CT scans, and data entry. The actual medical status of the patients who will receive help in medical care can be viewed using AI-based software. Technologies such as Blockchain will redefine the way electronic health records are managed and help connect them with other resources, such as payments and insurance. Also, in the case of any global pandemic like Covid-19, which has put immense pressure on the existing healthcare infrastructure, technologically enabled "Smart Cities" can respond in a much better and effective manner. Globally, there is an acute shortage of quarantining facilities for the patients, which coupled with fog of unawareness leaves the patients and healthcare professionals helpless. Imagine a scenario in which the "Bed Availability" of every hospital would be updated in real time in a central database and the same could be accessed by the citizens. Also, data related to life saving equipment such as PPEs, Ventilators, etc. was readily accessible to all the individuals involved in controlling the pandemic including forces, doctors, nursing staff, and citizens. All this not only would drastically reduce the response time but would also result in taking informed decisions at all levels.

8.6   COST OF LIVING AND SMART CITY

Cities can reduce transportation costs by improving the efficiency of train schedules and smart meters can reduce energy expenses at home and intelligent town-planned technology designed to address issues like air quality has a positive effect on healthcare costs. In order to facilitate quicker residential development and production, digitization of such procedures as the purchase of land, environmental analysis, and implementing it will theoretically minimize rental costs. In many Smart City projects, both the private and the public sectors benefit substantially and can generate increased consumer mobility and economic benefits. For these reasons, public-private partnerships can finance such projects (Tahir, 2017). During the last few decades, metropolitan areas around the world have engaged in initiatives to improve urban infrastructure and services for the improvement of environmental, social, and economic conditions (De Jong et al., 2015; Lee, Han, Leem, and Yigitcanlar, 2008). The concept of Smart City therefore has to necessarily incorporate and endeavour at enhancing the quality of life of its residents (Susanti, Soetomo, Buchori, and Brotosunaryo, 2016).

8.7   OPTIMIZED ENERGY CONSUMPTION

In 2016, the International Renewable Energy Agency (IRENA) report states that towns account for 65% of global consumption of energy and 70% of man-made carbon emissions (https://smartbuil-dingsmagazine.com/features/smart-cities-to-boost-energy-efficiency). The Bavarian village of Wildpoldsried, German population of around 2,600, produces 500% more renewable energy than it actually requires. Smart Nation Singapore mobile apps will allow citizens to monitor their electricity consumption and will propose ways to save energy. The Smart Town of Fujisawa in Japan connects 1,000 houses to a solar-powered intelligent grid just west of Tokyo and gives the area the ability to run off grid for up to 3 days (https://www.power-technology.com/features/smart-cities-redefining-urban-energy/). The city reports 70% less CO_2 emissions and returns 30% energy to the grid.

Over the last two decades, India has witnessed an unprecedented transformation from rural to the mainly urban landscape. In particular, the electricity supply is guaranteed with at least 10% of the energy demand for a "Smart City" being met from solar and intelligent metering, which must be included in the Smart City (Wendt, 2019). Other devices such as solar street lights, solar water heaters, solar pumps, and solar traffic signals as well as solar concentrator-based kitchen appliances can be exploited in addition to tapping solar energy on the rooftop for domestic consumption. The efficiency of solar production and intelligent storage solutions needs to be improved further in order to achieve these goals. Waste-to-energy is another technology of renewables that can contribute positively to this project. Solid waste produced can be converted into energy and other bioproducts which will make these intelligent cities self-sustaining as well as will meet the demands for electricity and organic fertilizer. In the planning of smart cities, biogas, wastewater treatment plants, and energy generation should be included. Around 30%–40% of conventional energy used in buildings may be saved by promoting and planning of green buildings in "Smart Cities" with integrated renewable power and energy conservation systems. Also, all the contributors of power generation have to be monitored by a "smart grid" which managed by AI, will then optimally utilize this power for the efficient functionality of a smart city (Figure 8.5.).

One-third of the world's energy consumption demand comes from buildings and this figure is expected to increase as urbanization progresses and cities expand. There is an urgent need for the right way forward, so that we conserve natural resources for our future generations. "Green Building" is the practice of creating structures and using processes that are environment-friendly and resource-efficient throughout a building's life cycle from siting to design, construction, operation, maintenance,

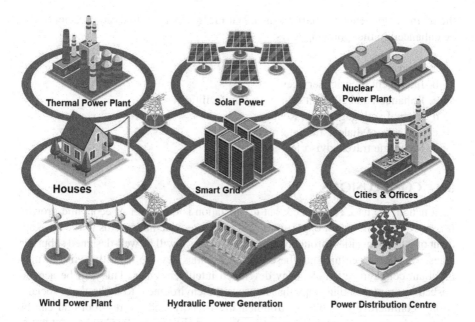

FIGURE 8.5 Smart grid solutions.

renovation, and deconstruction. These buildings, which will be the key dwelling components of any smart city, will have seven main components:

a. Energy Efficiency and use of Renewable Energy.
b. Efficient Water Management and Recycling Efficiency.
c. Environmentally Safe Building Materials and Specifications.
d. Waste Reduction.
e. Toxic Reduction.
f. Indoor Air Quality.
g. Smart Growth and Sustainable Development.

8.8 ENTERTAINMENT AND TOURISM IN A SMART CITY

Intelligent urban technology has the potential to produce a fully integrated events calendar with options for buying tickets, integrated with mobile payroll technology. This overlaps with "Intelligent Tourism" which generates data trails that "Smart Cities" capture and track as tourists utilize public transportation, hotels, restaurants, and other tourist attractions. Careful data analysis enables tourists in a "Smart City" to commute better, improve access to popular attractions, and continuously provide the update of packages or auxiliary services available.

With mobile access to information, tourists seek, find, and consume "Local experiences" with increasingly blurred limits between tourists and residents. Intelligent tourism includes a number of smart elements that are supported by ICT, such as shared economy, large-scale data applications, and industrial applications that are the main components of smart tourism. With IoT, smart city tourism experience can be enhanced in the following ways:

a. Smart tickets.
b. Intelligent security services.
c. Enhanced and efficient transportation facilities.
d. Virtual and Augmented Reality.
e. Services of dialect and interpretation.
f. Real time travel advisories.

8.9 CONCLUSIONS

As a nation, we have recognized that urbanization is the driver of economic growth. It is not a problem but an opportunity which will provide an impetus to the development of intelligent cities throughout the country, which will be well-designed suburban cities. We have to ensure that these intelligent cities develop from the existing metropolitan conglomeration to newly developed intelligent cities. This will be accomplished by implementing steps that improve the environment to make them socially strong, commercially viable, environmentally friendly, and future proof and can be achieved by means of urban regeneration. We need to focus on the Government initiative for Indian cities, the Smart City Mission, as an "Urban Renovation, Remedial, and Reconstruction Program" with a mission for developing smart cities across the country, which will make them "citizen friendly and sustainable". Sustainability can be implemented by reuse, recycling, and refilling. In relation to clever cities, sustainable projects ideally involve the implementation of the "Green Building" concept within an existing city. An eco-friendly and sustainable city also includes recycling waste into gardens, using methane produced from biodegradable waste for power generation, using wind and solar energy to supplement the energy requirements. We need to put dedicated efforts in areas such as air pollution control and water management, wastewater treatment, efficient connectivity for ensuring low pollution emissions on roads, e-governance, e-government system, and internet solutions, as well as the use of "eco-friendly" building materials which do not create an ecological imbalance. Thus, the "Smart City" initiative seeks to enhance the quality of life of people, develop and diversify the economy by prioritizing environmentally friendly solutions for its future.

REFERENCES

Aoun, C. *"The Smart City Cornerstone: Urban efficiency."* *Published by Schneider electric* (2013).

Berry, C. R., and E. L. Glaeser. "The divergence of human capital levels across cities." *Papers in Regional Science* 84, no. 3 (2005): 407–444.

Chandra, A., S. Moen, and C. Sellers. *What Role Does the Private Sector Have in Supporting Disaster Recovery, and What Challenges Does It Face in Doing So?* Santa Monica, CA: RAND Corporation, 2016.

De Jong, M., S. Joss, D. Schraven, C. Zhan, and M. Weijnen. "Sustainable–smart–resilient–low carbon–eco–knowledge cities; making sense of a multitude of concepts promoting sustainable urbanization." *Journal of Cleaner Production* 109 (2015): 25–38.

Fusco Girard, L. "Toward a smart sustainable development of port cities/areas: The role of the "Historic Urban Landscape" approach." *Sustainability* 5 no. 10 (2013): 4329–4348.

Giuffrè, T., S. M. Siniscalchi, and G. Tesoriere. "A novel architecture of parking management for smart cities." *Procedia-Social and Behavioral Sciences* 53 (2012): 16–28.

Gettinger, D. "Public safety drones: An update." *Center for the Study of the Drone* (2018): 1–14.

Gómez-Carmona, O., J. Sádaba, and D. Casado-Mansilla. "Enhancing street-level interactions in smart cities through interactive and modular furniture." *Journal of Ambient Intelligence and Humanized Computing* (2019): 1–14.

Glaeser, E. L., and C. R. Berry. "Why are smart places getting smarter." *Rappaport Institute/ Taubman Center Policy Brief* (2006): 2.

Joseph, T. "Smart city analysis using spatial data and predicting the sustainability." *arXiv preprint arXiv:1406.4986* (2014).

Khan, M. S., M. Woo, K. Nam, and P. K. Chathoth. "Smart city and smart tourism: A case of Dubai." *Sustainability* 9 no. 12 (2017): 2279.

Kumar, S., and M. Jailia. "Smart Cities with spatial data infrastructure and big data-a critical review." *International Journal of Advanced Studies of Scientific Research* 3 no. 11 (2018).

Lall, S. V. *Planning, Connecting, and Financing Cities—Now: Priorities for City Leaders.* Washington, DC: World Bank Publications, (2013).

Lazaroiu, G. C., and M. Roscia. "Definition methodology for the smart cities model." *Energy* 47, no. 1 (2012): 326–332.

Lee, S. H., J. H. Han, Y. T. Leem, and T. Yigitcanlar. "Towards ubiquitous city: Concept, planning, and experiences in the Republic of Korea." In *Knowledge-Based Urban Development: Planning and Applications in the Information Era*, pp. 148–170. Hershey, PA: IGI Global, (2008).

Li, W., M. Batty, and M. F. Goodchild. *"Real-Time GIS for Smart Cities."* (2020): 311–324.

Ministry of Urban Development. *"Smart Cities: Mission Statement and Guidelines."* Ministry of Urban Development, Govt. Of India, (2015).

Phadtare, S. M., and I. Jadhav. "Role of smart cities in sustainable development." *International Journal of Engineering Research and Technology,* 10 no. 1 (2017): 45–49.

Saeidi, S., and D. Oktay. "Diversity for better quality of community life: Evaluations in Famagusta neighbourhoods." *Procedia-Social and Behavioral Sciences* 35 (2012): 495–504.

Shah, S. A., D. Z. Seker, M. M. Rathore, S. Hameed, S. B. Yahia, and D. Draheim. "Towards disaster resilient smart cities: Can Internet of Things and big data analytics be the game changers?" *IEEE Access* 7 (2019): 91885–91903.

Susanti, R., S. Soetomo, I. Buchori, and P. M. Brotosunaryo. "Smart growth, smart city and density: In search of the appropriate indicator for residential density in Indonesia." *Procedia-Social and Behavioral Sciences* 227 no. 1 (2016): 194–201.

Tahir, M. S. "Public private partnerships (PPPs)." *The Professional Medical Journal* 24 no. 01 (2017): 1–9.

Wendt J. "Sustainability finds a home on the range." *The Solutions Journal*, 10 no. 1, Available at:(https://www.thesolutionsjournal.com/article/-sustainability-finds-home-range/), (2019, January).

Yigitcanlar, T., R. N. E. Hewa Heliyagoda Kankanamge, L. Butler, K. Vella, and K. Desouza. "Smart cities down under: Performance of Australian local government areas." (2020).

9 Integrating Human Well-Being and Quality of Life Derived from Nature-Based Livelihoods in Modern Economy

Parth Joshi
United Nations Development Programme

Aditi Joshi
Gradestack Learning Pvt. Ltd.

CONTENTS

9.1 INTRODUCTION

In his seminal work, *Zen and the Art of Motorcycle Maintenance*, American author and philosopher Robert M. Pirsig delves into the true meaning of the concept of 'quality' and postulates that it is not possible to define 'quality' since being inherently based on 'perception', it defies any efforts towards an 'intellectual' construction. However, he asserts, quality does exist and is in fact a fundamental force in the universe, and one can strive to understand it through analogies.

As human civilization has evolved into highly sophisticated social and economic systems, the definition of quality of life has also undergone changes, from the simple premise of longevity of life and abundance of resources into a complex concept deriving from an individual's own perceptions as well as external benchmarks established by the extant social structures.

How an individual person perceives quality to a large extent depends upon what we can call 'worldview'. A worldview is defined as 'inescapable, overarching systems of meaning and meaning-making that to a substantial extent inform how humans interpret, enact, and co-create reality' (Witt, 2013). We can classify worldviews into four distinct categories, as given in Figure 9.1 (Witt, 2014).

What individuals and societies in general view as quality of life differs according to these worldviews, as elucidated in Table 9.1 (Witt, 2014).

From the context of sustainable development, quality of life is perhaps the most important criteria to consider. If one looks at the Sustainable Development Goals (SDGs), adopted by United Nations Member States in 2015 as a successor to the

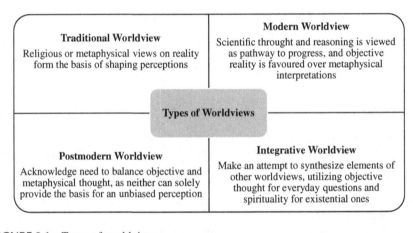

FIGURE 9.1 Types of worldviews.

TABLE 9.1

Quality of Life as Perceived According to Worldview

Type of Worldview	Perception of 'Quality of Life'
Traditional	• Focus on family and community bonhomie • Solicits adoption of traditional values like humility, sacrifice, obedience, loyalty and service
Modern	• Emphasis on 'individuality' and 'independence' • Promotes values of power, achievement, fame and hedonism
Postmodern	• Carving out a 'unique' individuality • Postmaterialist values like creativity, imagination and openness to change
Integrative	• An 'individuality' that acknowledges the interconnectedness of all life • Characterized by values such as coexistence, benevolence, service and self-actualization

FIGURE 9.2 The picture above depicts the SDGs which were set by the United Nations General Assembly in 2015 and which are a collection of 17 interlinked goals designed to be a 'blueprint to achieve a better and more sustainable future for all'.

Millennium Development Goals (MDGs), it becomes clear that the concept of 'quality of life' implies addressing multi-faceted and interconnected social, environmental and economic issues (Figure 9.2).

However, while on one hand, the SDGs chart an ambitious agenda for humanity's future, the seeming lack of clear pathways to achieve these targets remains an area of concern. As we discussed about the different types of worldviews earlier, the SDGs appear to derive mostly from a modern worldview, where economics is an enabler for achieving social and environmental goals rather than all three being placed on the same pedestal. While one can see some traces of an integrated worldview in the goals, there remains an ambiguity as to what does 'sustainability' exactly means, since it would have a varied definition and scope across different worldviews.

Still, the SDGs do provide a holistic framework that covers a majority of the indicators which can be used to evaluate or monitor quality of life. The Stockholm Resilience Centre has devised an interesting model to embed economic and social goals within goals pertaining to the biosphere (Stockholm University, n.d.). The approach, originally devised to highlight how food is inextricably linked to all SDGs, avoids the *faux pas* of falling into the trap of looking at economic, social and ecological goals in silos. This method underlines the importance, both intrinsic and extrinsic, of nature and natural ecosystems in our lives.

From a planning perspective, quality of life would ideally lie at the intersection of physical (including mental), societal and economic well-being. However, all these parameters would show a high degree of correlation. For instance, economic well-being would be determined by societal norms and expectations, while physical well-being would to some extent depend upon economic self-sufficiency. However, in terms of importance, it would be fair to say that physical well-being would be the topmost priority and the key parameter in evaluating quality of life (as goes the famous adage, 'Health is Wealth').

The World Health Organization defines Quality of Life as 'an individual's perception of their position in life in the context of the culture and value systems in which they live and in relation to their goals, expectations, standards and concerns' (World Health Organization, n.d.). Here, the concept of 'quality' is aligned towards ensuring good health including parameters like accessibility to healthcare (or universal healthcare from a policy standpoint) and the impact that an efficient healthcare system can have upon economic development and citizen welfare.

A more detailed approach to quality of life indicators is given by Schalock et al. through 24 indicators across eight domains that encapsulate the aforementioned three themes, as listed in Table 9.2. These indicators address both universal values (or 'etic') and values affiliated to culture (or 'emic'), thereby addressing both universal and relativistic aspects to determine quality of life (Schalock et al., 2005).

The Human Development Index (HDI) is another approach to assess the development of a country beyond economic growth. Built across three major criteria,

TABLE 9.2
Quality of Life Domains and Indicators (Schalock et al., 2005)

Domain	Indicators and Descriptors
Emotional well-being	• Contentment (satisfaction, moods, enjoyment) • Self-concept (identity, self-worth, self-esteem) • Levels of stress
Interpersonal relations	• Interactions (social networks) • Relationships (family, friends, peers) • Support (emotional, financial, physical)
Material well-being	• Financial status • Employment • Housing
Personal development	• Education and skills • Competence, professional ability • Performance, productivity
Physical well-being	• Health • Lifestyle • Access to healthcare • Leisure activities
Self-determination	• Autonomy, independence • Goals, desires • Opportunities, choices
Social inclusion	• Community participation • Community roles • Social support system
Rights	• Human rights • Legal rights

viz. Life expectancy, Knowledge (or education) and Standard of living (calculated through Gross National Income). The development of the index was motivated to provide a simple yet all-encompassing metric to shift the single-pointed focus from economic growth and output to a more people-centric approach.

Many economists utilized this approach to broad-base interpretation of development including Nobel laureate Amartya Sen who used this to develop his human capabilities approach which emphasizes the importance of ends (like a decent standard of living) over means (like income per capita) (Stanton, 2007).

However, one must also acknowledge that these pillars and parameters thought to attribute a measurability to ascertain quality of life are not without their own contradictions. For instance, one might amass enormous wealth but might suffer from ailments that hamper enjoyment of life, or an individual might have very limited independence or autonomy yet might be completely satiated.

There exist many such conundrums that resist the utilitarian approach, which developed in early 19th century advocated the use of a single parameter to define human well-being. While John Stuart Mill rectified this to a certain extent by proposing a range of qualities arranged in a hierarchy and recognizing the influence of social factors upon determining well-being in the latter half of 19th century (Mill, 1870), the model still remained inherent flawed.

This brought about the development of HDI more than a century later, preceded by the evolution of the marginalist and ordinalist schools of thought before the Second World War and eventually the humanist school of thought in the 1970s. While HDI tried to imbibe a pluralistic approach to defining the quality of life and thus, the larger premise of sustainable development.

However, we must note that all these developments were approached from the points on economists, and no matter how inclusive the models were, eventually ended up centred around Gross Domestic Product (GDP) and economic development as the sole means to achieve well-being in all other spheres.

In all these approaches, we observe that the role of nature as a conduit towards ensuring quality of life is rather restricted, especially when consider a micro, individualistic approach. While one can point out that it is not an easy task to quantify the role of something as innate as nature when we consider human well-being, its role remains highly understated.

When we consider that increased intervention of technology in our daily lives in the past few decades has completely revolutionized how economies work and dramatically increased the process of urbanization, instance of lifestyle diseases that were hitherto unheard of have also started increasing proportionally.

While there is a need to conduct a lot more research to quantify or accurately ascertain how this model of modernization, which has restricted our interactions with natural ecosystems, might be responsible for this, there is a substantial body of evidence that clearly points in this direction.

In this chapter, we take a look at three nature-based livelihoods, *viz.* agriculture, pastoralism and high-altitude mountain porters or *Sherpas*, and try to understand why they offer a better quality of life and sense of fulfilment. Based on this, we propose a framework through which the values in these livelihoods can be integrated into modern economies, especially in urban environments dominated by the services sector.

9.2 MEASURING NATURE'S CONTRIBUTION TO HUMAN WELL-BEING AND QUALITY OF LIFE AT A MACRO-LEVEL

Human beings rely upon nature for survival, either directly or indirectly. We derive the fundamental ingredients of life, *viz.* air, water, food from them, as well as other resources essential to maintain physical health as well as socio-economic structures. Even today, despite the exponential growth of the services sector, economies are primarily valued on the amount of natural resources a region or a country possesses.

As individuals, we have an innate habit of seeking connection with nature and other forms of life. The Biophilia Hypothesis, propounded by American biologist E.O. Wilson, asserts that as a part of our evolutionary process, the tendency to seek connection with nature and other life forms is built into our genetic framework (Wilson, 1984). This is apparent if we look at cultural memes, for example, where phrases like 'dead as a dodo' or 'like a sitting duck' that invoke nature in language are quite common across many languages.

By weaving nature inextricably into the socio-cultural fabric, our forefathers acknowledged the fact that 'we need nature, nature doesn't need us'. By deifying pristine ecosystems, for instance forest groves, rivers and mountains, it was ensured that these ecosystems were revered for the intangible values they provide and not harnessed unsustainably for resources. Countless animistic traditions can still be found across many cultures even today that hold testament to this fact.

What happens when we lose our connect with nature? To put it simply, our growth as healthy and happy individuals is negatively affected. Right from childhood, nature plays an important part in our development, both in terms of cognitive as well as emotional abilities. Alienation of children from natural surroundings in recent decades is already showing adverse impacts, from rising physical ailments and lifestyle diseases to stunted physical and mental growth. As they grow up, this would have a direct impact upon the productivity of the workforce and thereby economic growth.

On the other hand, loss of children's contact with the natural world also sets the stage for a continuing loss of the natural environment. The alternative to future generations who value nature is an apathetic society that continues the exploitation and destruction of nature as we see today. Research has clearly demonstrated that an affinity to and love of nature, along with a positive environmental ethic, grow out of children's regular contact with and play in the natural world (Chawla, 2013).

At the other end of the spectrum, connection with nature is equally important for old-aged persons to ensure their health and well-being. While reduced mobility limits nature experiences especially for frailer adults, in contrast, for 'younger' older adults, retirement can provide more time and opportunities to engage with nature. Nature connection opportunities are valued by nearly all adults irrespective of age and health (Freeman et al., 2019).

Moving away from the individual perspective to consider our connection with nature at a macro level in terms of societies and economies, we arrive at the concept of ecosystem services which elucidates the tangible values and commodities that nature provides us. This concept was popularized by the Millennium

Ecosystem Assessment (MA) supported by the United Nations that undertook a large-scale assessment of human impact upon the environment. As per this framework, ecosystem services can be grouped into four categories, as depicted in Figure 9.3 below.

Building upon this, it is important to take into account that people and societies might have very different perceptions of how nature contributes to their quality of life and well-being, depending upon factors such as economic status and cultural affiliations. Consider the *Chipko Andolan* in the 1970s that originated in the mountain regions of the Indian state of Uttarakhand (then Uttar Pradesh) where villagers with relatively modest means staged a non-violent protest against indiscriminate felling of trees (Bhatt, 1990).

While this movement was embedded in people's awareness of why protection of forests was important for their economic well-being, the precursor to this was a movement in the state of Rajasthan (as a desert an entirely different ecosystem to the Himalaya), more than two centuries before in 1730 AD where nearly 360 people of Bishnoi tribe were massacred by the local king's soldiers when they protested felling of *Khejri* trees (*Prosopis cineraria*) as it was against their religious beliefs (Panwar et al., 2014).

The management and resolution of local or global ecological distribution conflicts requires cooperation between business, international organizations, NGO, community groups and governments. But it is difficult for such a cooperation to be based on common values as even perceptions on economic evaluations might differ. Some people want to conserve nature because they derive their subsistence from them, while some might argue that there is a need to harness and not protect them for the greater good (Martinez-Alier, 2002). Therefore, the yardstick of economic evaluation may not be the best tool to ascertain the 'value' of nature and its importance in ensuring quality of life.

The Economics of Ecosystems and Biodiversity (TEEB) was one such initiative launched in 2007 after a meeting on the environment to ministers of G8+5 to

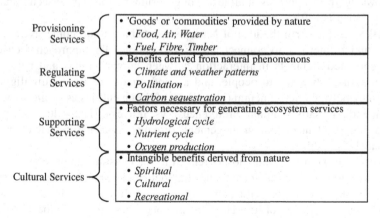

FIGURE 9.3 Categories of ecosystem services. (Adapted from Millennium Ecosystem Assessment).

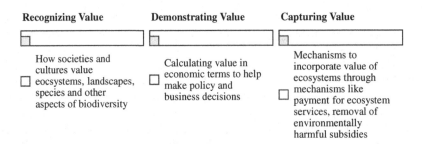

FIGURE 9.4   Core principles of the TEEB study (Sukhdev et al., 2010).

ascertain the costs of loss of biodiversity and compare the costs of increasing conservation measures vis-à-vis maintaining status quo (The Initiative, n.d.).The study
followed three core principles, as given in Figure 9.4.

However, the report also acknowledges the fact that the natural science underpinning many economic valuations remains poorly understood, and both economics and
ethics demand more systematic attention to the values of biodiversity and ecosystem
services. Another critique of the methodology is the use of human preferences as
a parameter, which is rather controversial as humans being are a 'part' of the ecosystem and therefore it is questionable whether they have the right or capability to
determine what should be constituents of a healthy ecosystem.

Recent years have seen an increased impetus upon taking a more holistic and
inclusive approach to valuing nature and ecosystem services and create mechanisms
that acknowledge its non-material, non-economic contributions to ensure quality of
life as against the 'commodification' that has inevitably become the norm in modern
economic structures.

The Intergovernmental Science-Policy Platform on Biodiversity and Ecosystem
Services (IPBES) is an intergovernmental organization established in 2012 with 134
member states to provide policy-relevant knowledge and catalyse the implementation
of knowledge-based policies at all levels in government, the private sector and civil
society (What Is IPBES, n.d.).

IPBES has put forth the idea of Nature's Contribution to People (NCP) as the
conceptual framework to connect nature and people. The approach focuses on
three corelated elements in interaction between human beings and nature, *viz.*
nature, nature's benefits to people, and a good quality of life. It highlights the
dichotomy between instrumental (i.e., values of living entities as means to achieve
human ends, or satisfy human preferences), vs. intrinsic (i.e., values inherent to
nature, independent of human judgement) dimensions of nature (Heather Tallis
and Lubchenco, 2014).

Some NCP parameters are closely related to the constituents of a 'good quality
of life', the aspects of a relational values that define elements like cultural identity,
social cohesion and responsibility, and moral responsibility towards nature, and can
impinge on good quality of life. This makes it necessary to expand the way society
recognizes the diversity of values and to embrace pluralistic valuation approaches
(Pascual et al., 2017).

9.3 LOOKING AT NATURE-BASED LIVELIHOODS TO UNDERSTAND QUALITY OF LIFE

All of these above approaches are aligned towards impacting macro-policies and mechanisms to urgently mitigate the widespread commodification of nature, so no matter how inclusive the approach might be towards including non-tangible, relational values, the element of 'valuation' eventually tends to dominate and subdue the envisaged impact of subjective interpretations.

A more 'bottom-up' approach to looking at quality of life could be to look at specific nature-based livelihoods, look at the quality of life of the practitioners, and extract values from them which makes these occupations enriching and fulfilling, and try to integrate them into systems where quality of life seems to be suffering due to lack of interaction with nature.

We analyse the quality of life in three occupations, *viz.* agriculture, pastoralism and mountain porters. While the first two are 'old-world' professions, the third is of recent origins with the advent of adventure tourism and mountaineering as an industry. All of these involve a high degree of interaction with natural environments.

9.3.1 AGRICULTURE

One of the oldest occupations that led to the rise of sedentary human civilization, one can say that agriculture has, by allowing a large part of the population to undertake other tasks than growing or foraging for food, shaped the human world as we see it today.

While in a certain context, farming today is a sector plagued with many issues, especially for traditional and smallholder farmers in the 'global south' who are struggling to make ends meet and survive in the profession, for the purpose of this chapter, we will restrict our focus on farmers with a sufficient economic means and support infrastructure.

Quality of life is the key driver for rural development activities. Rural residents, especially farmers, in the United States have been found to express higher ratings of quality of life than their urban counterparts, especially with regard to their physical and social surroundings (Campbell, 1981), and economic development strategies are primarily based on methods through which rural areas can capitalize on this perceived advantage (Morgan et al., 2009). It has been found that quality of life among farm families is tied more strongly to the non-farming elements of their lives including abundant natural amenities and cohesive social structures (Arbuckle and Kast, 2007).

While economic incentives are being provided to farmers in developed nations, especially Europe, to undertake conservation practices, studies have shown that many farmers are already undertaking or are willing to undertake conservation practices without and government incentives, with the concepts of self-identity and personal norms appear to be related to the intention to perform non-subsidised, conservation (Lokhorst et al., 2011).

In the UK, it has been found that policy knowledge cultures, which are accommodative of farmers' ways of knowing nature are more effective tools for conservation than codified knowledge of agri-environment schemes which are top-driven by the government (Morris, 2006). In Australia, a research found that 'connected to nature', or CNS, which is a scale to measure an individual's emotional connect to measure, was a more effective factor in protecting native management than the sense of ownership of their farm (Gosling and Williams, 2010).

It is also interesting to note that the utilization of agricultural farms as a factor to improve human mental and physical health is on the rise in several developed countries. This concept, known as *Green care*, was earlier restricted to hospitals and psychiatric institutions (Berget et al., 2008) but in recent years is being promoted as a general well-being tool in the form of agri-tourism and farm tours.

In Sweden, it was found that farmers and nonfarming rural men had significantly lower morbidity risks of cancer and psychiatric disorders, as well as endocrine disorders, cardiovascular disorders, and respiratory disorders than urban referents (Thelin et al., 2009). Even within the same rural area in Greece, lower prevalence of hypertension was recorded in young farmers in relation to young non-farmers, although this trend did reverse with older ages (Demos et al., 2013).

We surmise here that working in a predominantly outdoors environment leads to an overall better level of physical and mental health, and a sense of fulfilment is derived from living in social structures with community support.

We must note that farmers across the world face many stresses including the vagaries of weather, unfair market dynamics, financial strains etc., which in recent years have led to increasing instances of suicides. However, the primary reason for that is attributable to financial matters and not to any natural factors. In traditional farming systems, loss of a cycle or two of crops did not place farmers under much duress, but this has been a failing of the policy mechanism as a whole in not providing farmers with appropriate safety nets.

In a nutshell, we can say the farmers derive the following values from working in a natural environment:

- **Gratitude** towards nature for providing livelihood
- **Compassion** towards farm animals that provide them with meat, manure and milk
- **Tolerance** to bear losses due to many uncertainties in the profession
- **Empathy** towards fellow human beings from living in close-knit communities
- **Satisfaction** from living a life in clean, unpolluted environment

9.3.2 PASTORALISM

Like agriculture, pastoralism is one of the oldest occupations in human history, and both are believed to have originated side by side as an evolution from hunting and gathering. As compared to farming, pastoralism is more dependent upon wild, 'as they are' ecosystems, and instead of modifying it to suit one's needs, relies on interacting with the environment to convert one form of food into another, i.e., plant to meat or dairy, and if we consider the case of ungulates, from food to fibre.

As pastoralists need to move large distances and mobility between political borders as well as access to natural ecosystems is becoming difficult by the day, the occupation is on the decline. In developed countries, nomadic pastoralism is almost over and has given way to sedentary pastoralism, but it still remains a way of life in many developing nations, especially those living in challenging environmental conditions such as extreme temperatures, aridity and low soil productivity which renders agricultural activity unproductive.

Nomadic pastoralists have a deep connection with their natural surroundings, as they often move hundreds of kilometres in difficult terrains for months with minimal human contact, often at the mercy of natural forces. Although they do not generally travel alone, due to large livestock sizes that need sizeable tract of land, they seldom move in large groups.

One finds a high degree of deification, especially of inanimate natural elements like mountains, rivers or oases along pastoral routes, highlighting the fact that traditional belief systems connecting humans with nature help in maintaining the mental well-being in the absence of family or community support in the wilderness.

From the perspective of natural anthropology, nomadic pastoralism throws up some very interesting insights. To move within large areas in harsh landscapes that might seem a daunting task, the psychology of a pastoralist tries to comprehend these large 'spaces' as a composition of 'places', for instance shepherd camps along migration routes. Despite only traversing a small percentage of the total landscape areas, a pastoralist has to make himself aware of the 'lay of the land', initially through traditional knowledge and then through personal experience (Wagner, 2013).

We consider the case of *Gaddi* nomadic pastoralists, native to the state of Himachal Pradesh in India. Traversing high-altitude Himalayan landscapes, they lead a semi-nomadic lifestyle, migrating with their livestock to high altitude meadows and pastures in the summers and spending the winters at home, an activity which is often considered as the largest scale sheep and goat herding in the entire Himalayan region (Tucker, 1986). Transhumance also implies a greater degree of women empowerment, since men are out of the house for a large part of the year, women enjoy a greater say in family life and a higher degree of social freedom (Pandey, 2011).

Moving in some of the most biodiversity-rich regions, the *Gaddis* are known for their ethnobotanical knowledge and wisdom, using over 450 plant species and products for food, fibre, socio-religious ceremonies, medicine, veterinary medicine, oil, gum and resin, dye and tannin, tribal crafts, timber and woodwork, fuelwood, fodder and narcotic drinks (Singh and Kumar, 2000).

Economically, the *Gaddis* earn a decent living. A study conducted on registered *Gaddi* shepherds in Himachal Pradesh revealed that their average income increased from INR 5.2 lakhs in 2010–2011 to INR 13.04 lakhs in 2016–3017 (Dogra et al., 2018), which makes it a viable rural enterprise. Looking at their modest lifestyles though, it is difficult to gauge that they earn such amounts.

Such quantum of earnings allows them the opportunity to diversify their income-generating activities and reduce dependency on pastoralism, which remains an inherently risky profession as they move in remote high-altitude areas prone to weather extremes and natural calamities. However, the fact that they keep up with this activity implies that this is more than just a profession, and rather a way of life for these pastoralists.

This in part can be attributed to their deep affinity with natural surroundings, where they spend months roaming in pristine wilderness. They breathe some of the purest air and drink the cleanest water, and undertake lots of physical activity which keeps them in prime health. Interaction with nature for prolonged periods also ensures their mental well-being.

In Africa, research has shown that there is a correlation between environment and mental health in pastoral communities, and any concern over the degradation of natural habitats or scarcity on natural resources has a negative impact upon their mental well-being (Cooper et al., 2019). Links between environmental conditions and an affective relationship with nature is an emerging area of research (Marczak and Sorokowski, 2018) that can give us more insights on how the natural environment, including its vagaries, impact the well-being of pastoralists.

It can be surmised that prolonged interaction with nature inculcates the following values in pastoralists:

- **Gratitude** towards nature for providing livelihood
- **Compassion** towards their herd
- **Animism**, ascribing 'life' or 'soul' to plants, inanimate objects, and natural phenomena
- **Humility** to live a modest lifestyle
- **Empathy** towards fellow human beings
- **Satisfaction** from living a life in clean, unpolluted environment
- **Idyll,** a longing for spending time in solace with nature

9.3.3   *Sherpas* (High-Altitude Mountain Porters)

Unlike agriculture and pastoralism, mountain porters are a relatively recent phenomenon, coming to the fore as mountaineering, especially in the Himalayan region evolved from a colonial quest to a thriving industry in the span of a century and a half. While the occupation can trace its roots to the time of slavery when there weren't many ethical issues about using human beings as beasts of burden, today it exists in regions where transportation of goods through mechanical conveyance is not possible.

Today, the *Sherpa* community of Nepal constitutes some of the most skilled mountain porters, with their abilities going beyond simple load bearing to the technical nuances of running mountain expeditions, so much so that *Sherpa* is now a commonly accepted term for referring to high-altitude porters. Due to this, they are generally better paid than other porters and are also treated with respect as the success of expeditions literally rests on their shoulders.

On an average, a *Sherpa* earns between US\$ 3,000–10,000 in a climbing season, which is much higher than the national average of US\$ 700 per annum (K.S.C., 2015). There are two factors to consider here though. Firstly, at 1.2%, there is a high risk of mortality in the job which exists in no other service industry in the world (Schaffer, 2013). Secondly, a western guide, who climbs with paying clients and undertakes much less work, on an average gets paid US\$ 50,000. From this perspective, the risk-reward ratio for a *Sherpa* doesn't look good.

Sherpa Buddhism is an amalgamation of local shamanism with more formal Tibetan Buddhism, replete with animistic traditions implying a high degree of deification to their natural surroundings, with concepts of sacred landscapes and natural features. These deities must be pleased for the residents to be protected from natural disasters like avalanches, landslides and flash floods.

While the culture of abstaining from summitting mountains, supposed to be the abode of gods, has somehow waned as western expeditions were doing it anyway, summits are often adorned with Tibetan prayer flags, with Sherpas offering their prayers on each successful summit. Despite modernization, the connection of *Sherpas* to the traditional values has remained strong probably due to the precarious nature of mountaineering. In fact, modernization, by which we refer to the commodification of the *Sherpa* way of life as a tourism product may have even helped in conservation of these traditional values (Spoon, 2012).

When we look at *Sherpa's* connection with nature, there is a balance between traditional Buddhist beliefs that advocate deep ecology and the Animistic or Shamanistic traditions that harness nature for their own gains without much regard for conservation. The concept of 'pollution' is restricted to human nature (needs, behaviour and desires) and has little to do with environmental pollution (Obadia, 2008). In this scenario, it is interesting to note that modernization in the sense of tourism development and the subsequent environmental regulations put in place to mitigate harmful impacts upon the environment have sensitized *Sherpas* towards the issue and promoted environmental stewardship.

From anecdotal impressions, *Sherpas* in general are perceived as a tranquil and happy people with simple, quite lives, a broad network of family and friends, and a contentment from living in one of the most beautiful landscapes on the planet (Morgan, 2019). Their physical capabilities from living in extreme environments has provided them with better livelihood opportunities without the need for migrating to urban or low-lying areas. In recent years, *Sherpas* have emerged as successful professional mountaineers themselves and carved out a niche for themselves in extreme adventure sports.

We can summarize the following values that *Sherpas* derive from working in the great outdoors:

- **Determination** to undertake some of the most challenging and risky physical tasks
- **Respect** for nature, both through traditions and in practice due to the dangerous nature of their profession
- **Preserving** traditional values in the face of modernization
- **Perspective** to life from living in overwhelming landscapes and constantly facing mortal risks

9.4 IMBIBING VALUES DERIVED FROM NATURE-BASED LIVELIHOODS IN OTHER SECTORS

While nature-based livelihoods come with their own set of challenges and risks to physical and mental health, they still provide for a healthier and more fulfilling work environment where one can take a more holistic view of concepts like success, wealth and achievement, and avoid falling into siloed benchmarks of market driven thought process that can negatively impact physical as well as mental well-being.

Three key values derived from nature-based livelihoods are proposed which if imbibed into other sectors of modern socio-economic structures can ensure physical as well as mental well-being and contribute tangibly towards improving quality of life.

9.4.1 DYNAMISM OVER STABILITY

Nature is in a constant state of flux, and practitioners of nature-based livelihoods are attuned to this fact, always aware of the environment around them and constantly improvising to ever-changing scenarios. While the idea of 'change is permanent' is well entrenched in large organizations and economies who are always planning towards it, at the individual level, 'stability' is the norm, putting onus on individuals to choose a path, be it career or family, and stick with it no matter how the circumstances around them change.

This rigid approach can eventually cause stress, both physical and mental, and negatively impact quality of life. To avoid this, both our social and economic frameworks must give individuals 'flexibility', in terms of how they set goals and expectations and how they go about achieving them. For instance, as the COVID-19 pandemic has shown, work-from-home option for a large part of the workforce, which would never have been thought of previously, is now the norm, and offers an opportunity to how we define work-life balance in the future.

The Gross National Happiness concept adopted by the country of Bhutan tries to redefine the meaning of quality of life by broad basing the parameters for well-being instead of just focusing on economic prosperity and access to modern amenities. Integrating such philosophies into each tier of governance system will go a long way in widening the currently myopic view of human well-being, and make us much more receptive to change.

9.4.2 RETHINKING PRODUCTIVITY

The western school of thought has traditionally regarded work as synonymous to attaining freedom, both personal and economic. Time spent in idyll is considered as waste, so we try to maximize the amount of time working and minimize the time spent doing nothing to increase the productivity of the workforce.

However, we are at a juncture today when modern-day technologies are capable of handling a majority of the tasks requiring human intervention, and this list is growing exponentially by the day. In such a scenario, the current paradigms of productivity, if unchanged, might lead to desperation and depression in the working class.

This needs a complete shift in the current thought structures, and we should not be looking being idle as the opposite of empowerment or independence. Pastoralists or porters spend a lot of time sitting idle, doing nothing but stare at vistas in front of them, and what may look like rest is actually time spent in self-contemplation and evaluation, an exercise which many people today forego and choose to adopt the readymade standards hard coded into socio-economic structures, which can eventually lead to dissatisfaction or a sense of disillusionment.

9.4.3 BEING 'WILD', NOT 'MANICURED'

When we look at the example of green spaces in well-planned cities, the idea of trimmed trees and hedges and manicured parks come to mind. However, nature thrives in chaos, not order. Consider the example of Singapore, one of the most well-planned cities in the world. During the COVID-19 lockdown, the maintenance of green spaces was considered non-essential, due to which the otherwise manicured spaces were allowed to grow wild. This led to an unprecedented increase in biodiversity, with more plants, butterflies and birds being observed (Asher, 2020).

Pastoralists always prefer to take their herds out to the wild for grazing over stall feeding. Horticulturalists leave wild spaces to attract pollinators. Our quality of life often suffers from being 'manicured', from creating rigid work or social structures adhering to which can take away the creativity that thrives in chaos and negatively affect personal growth.

As the innovation culture is steadily growing, this approach is seeing a revival with many organizations 'redefining the workplace', encouraging employees to 'think outside the box' among others, leading to improvement in employee satisfaction. This approach needs to be adopted in a more widespread manner, for instance in government systems, where a focus on innovation can overhaul archaic systems.

9.5 CONCLUSIONS

Human civilization is standing at the crossroads where rapid advances in technology have greatly improved the quality of life in respect of safety, security and physical well-being. However, recent decades have also seen a sharp increase in the frequency of lifestyle diseases that include mental illnesses, raising concerns on whether the perceived notions of comfort and security are effective parameters for determining the quality of life.

It has been observed that even today, livelihoods and occupations pertaining to primary sectors like agriculture and those directly dependent upon natural ecosystems and resources like pastoralism, cultivation and collection of medicinal plants etc., despite their low remunerative capacities, provide a better quality of life for people engaged in these activities in terms of their physical and mental health, as well as a better social support system through more cohesive and close-knit communities.

With most livelihoods in modern economic frameworks moving towards the services and manufacturing sectors where interaction with nature is minimal, organizations are looking for ways to ensure well-being of the workforce. Therefore, it becomes important to assess the values that nature-based livelihoods provide which ensure a better quality of life for humans and create mechanisms that enable us to transition these values into modern-day livelihoods.

BIBLIOGRAPHY

Arbuckle Jr, J.G., and Kast, C (2007). Quality of life on the agricultural treadmill: Individual and community determinants of farm family well-being. *Journal of Rural Social Sciences*, 84–113.

Asher, S. (2020, June 14). *Coronavirus in Singapore: The Garden City Learning to Love the Wild*. Retrieved from BBC News: https://www.bbc.com/news/world-asia-52960623.

Berget, B., Ekeberg, Ø., and Braastad, B.O. (2008). Animal-assisted therapy with farm animals for persons with psychiatric disorders: effects on self-efficacy, coping ability and quality of life, a randomized controlled trial. *Clinical Practice and Epidemiology in Mental Health*, 4: 1–7.

Bhatt, C.P. (1990). The Chipko Andolan: Forest conservation based on people's power. *Environment and Urbanization*, 2: 7–18.

Campbell, A. (1981). *The Sense of Well-Being in America: Recent Patterns and Trends*. New York: McGraw-Hill.

Chawla, L. (2013). Bonding with the natural world: The roots of environmental awareness. *The NAMTA Journal*, 2: 39–51.

Cooper, S., Hutchings, P., Butterworth, J., Joseph, S., Kebede, A., Parker, A., Terefe, B., and Van Koppen, B. (2019). Environmental associated emotional distress and the dangers of climate change for pastoralist mental health. *Global Environmental Change*, 59: 101994.

Dogra, P., Sankhyan, V., Kumari, A., Thakur, A., and Kumar, N. (2018). Enhancing profitability of nomadic Gaddi goat production system for augmenting farmer's income. *Indian Journal of Animal Production and Management*, 34: 36–39.

Demos, K., Sazakli, E., Jelastopulu, E., Charokopos, N., Ellul, J., and Leotsinidis, M. (2013). Does farming have an effect on health status? A comparison study in west Greece. *Environmental Research and Public Health*, 776–792.

Freeman, C., Waters, D.L., Buttery, Y., and van Heezik, Y. (2019). The impacts of ageing on connection to nature: The varied responses of older adults. *Health & Place*, 56: 24–33.

Gosling, E., and Williams, K.J. (2010). Connectedness to nature, place attachment and conservation behaviour: Testing connectedness theory among farmers. *Journal of Environmental Psychology*, 30: 298–304.

K.S.C. (2015, December 11). *The price the Sherpas pay for Westerners to climb Everest*. Retrieved from The Economist: https://www.economist.com/prospero/2015/12/11/the-price-the-sherpas-pay-for-westerners-to-climb-everest.

Lokhorst, A.M., Staats, H., van Dijk, J., van Dijk, E., and de Snoo, G. (2011). What's in it for Me? Motivational differences between Farmers' subsidised and non-subsidised conservation practices. *Applied Psychology*, 60: 337–353.

Marczak, M., and Sorokowski, P. (2018). Emotional connectedness to nature is meaningfully related to modernization. evidence from the Meru of Kenya. *Forntiers in Psychology*, 9: 1789.

Martinez-Alier, J. (2002). *The Environmentalism of the Poor*. Geneva: United Nations Research Institute for Social Development (UNRISD).

Mill, J. S. (1870). *Utilitarianism*. London: Longmans, Green and Company.

Morgan, N. (2019). *In the Mountains: The Health and Wellbeing Benefits of Spending Time at Altitude*. London: Hachette UK.

Morgan, J.Q., Lambe, W., and Freyer, A. (2009). Homegrown responses to economic uncertainty in rural America. *Political Science*: 15–28.

Morris, C. (2006). Negotiating the boundary between state-led and farmer approaches to knowing nature: An analysis of UK agri-environment schemes. *Geoforum*, 37: 113–127.

Obadia, L. (2008). The conflicting relationships of Sherpas to nature: Indigenous or western ecology? *Journal for the Study of Religion Nature and Culture*: 116–134.

Pandey, K. (2011). Socio-economic status of tribal women: A study of a transhumant Gaddi population of Bharmour, Himachal Pradesh, India. *International Journal of Sociology and Anthropology*, 3: 189–198.

Pascual, U., Balvanera, P., Díaz, S., Pataki, G., Roth, E., Stenseke, M., Watson, R.T., Dessane, E.B., Islar, M., Kelemen, E., and Maris, V. (2017). Valuing nature's contributions to people: The IPBES approach. *Environmental Sustainability*, 26: 7–16.

Panwar, D., Pareek, K., and Bharti, C.S (2014). Unripe Pods of Prosopis cineraria used as a vegetable (sangri) in Shekhawati region. *International Journal of Scientific & Engineering Research*: 892–895.

Schaffer, G. (2013, July 10). *The Disposable Man: A Western History of Sherpas on Everest.* Retrieved from Outside: https://www.outsideonline.com/1928326/disposable-man-western-history-sherpas-everest.

Schalock, R.L., Verdugo, M.A., Jenaro, C., Wang, M., Wehmeyer, M., Jiancheng, X., and Lachapelle, Y. (2005). Cross-cultural study of quality of life indicators. *American Journal on Intellectual and Development Disabilities*, 110: 298–311.

Singh, K.K., and Kumar, K. (2000). *Ethnobotanical Wisdom of Gaddi Tribe in Western Himalaya.* Dehradun: Bishen Singh Mahendra Pal Singh.

Spoon, J. (2012). Tourism, persistence, and change: Sherpa spirituality and place in Sagarmatha (Mount Everest) national park and buffer zone, Nepal. *Journal of Ecological Anthropology*, 15: 41–57.

Stanton, E.A. (2007). *The Human Development Index: A History.* Amherst: Department of Economics, University of Massachusetts-Amherst.

Stockholm University. (n.d.). How food connects all the SDGs. Retrieved from Stockholm Resilience Centre: https://www.stockholmresilience.org/research/research-news/2016-06-14-how-food-connects-all-the-sdgs.html.

Sukhdev, P., Wittmer, H., Schröter-Schlaack, C., Nesshöver, C., Bishop, J., Brink, P.T., Gundimeda, H., Kumar, P., and Simmons, B. (2010). *The Economics of Ecosystems and Biodiversity: Mainstreaming the Economics of Nature: A Synthesis of the Approach, Conclusions and Recommendations of TEEB.* United Nations Environment Programme.

Tallis, H., and Lubchenco, J. (2014). Working together: A call for inclusive conservation. *Nature News*, 515: 27–28.

The Initiative. (n.d.). Retrieved from The Economics of Ecosystems & Biodiversity: http://www.teebweb.org/about/the-initiative/.

Thelin, N., Holmberg, S., Nettelbladt, P., and Thelin, A. (2009). Mortality and morbidity among farmers, nonfarming rural men, and urban referents: A prospective population-based study. *International Journal of Occupational and Environmental Health*, 15: 21–28.

Tucker, R.P. (1986). The evolution of transhumant grazing in the Punjab Himalaya. *Mountain Research and Development*: 17–28.

Wagner, A. (2013). *The Gaddi beyond Pastoralism: Making Place in the Indian Himalayas.* New York: Berghahn Books.

What Is IPBES. (n.d.). Retrieved from ipbes: https://ipbes.net/about.

Wilson, E.O. (1984). *Biophilia.* Cambridge: Harvard University Press.

Witt, A.H.-d. (2013). Worldviews and their significance for the global sustainable development debate. *Environmental Ethics*, 35: 133–162.

Witt, A.H.-d. (2014). Rethinking sustainable development: Considering how different worldviews envision 'Development' and 'Quality of Life'. *Sustainability*, 6: 8310–8328.

World Health Organization. (n.d.). *WHOQOL: Measuring Quality of Life.* Retrieved from World Health Organization: https://www.who.int/healthinfo/survey/whoqol-qualityoflife/en/.

10 Nurturing Resilience and Quality of Life
A Blue Economy Approach

Belinda Bramley
NLA International

Angelique Pouponneau
Seychelles' Conservation and Climate
Adaptation Trust (SeyCCAT)

Upeksha Hettiarachchi
Duke University

CONTENTS

10.1 INTRODUCTION

At the start of the new decade, a clear theme has surfaced: resilience has become the new watchword for 2020 and beyond. How do countries build back better in challenging times and restore the health of people and the natural world, reduce carbon emissions and provide decent work to deliver long-term inclusive and sustainable well-being and prosperity? Economic and social vulnerabilities have been cruelly exposed by the COVID-19 pandemic of 2020 and fuelled a growing recognition that a quantum shift is needed in the way we perceive and create economies and the values they bring.

We live on a blue planet, and in recent years, attention has been shifting to the ocean for the wealth of possibilities on offer. The world's ocean comprises the greatest ecosystem on earth, supporting all life above and below water. It provides 99% of the living space by volume on earth and half of the oxygen we breathe. More than half of the world's population live in coastal zones. If the ocean was a country, it would have the seventh largest economy in the world, with an annual gross marine product in excess of US$2.5 trillion (Hoegh-Guldberg et al., 2015).

The ocean has a deep and fundamental connection with our quality of life. In addition to providing a liveable climate, food and livelihoods, marine and coastal landscapes provide therapeutic effects which play a significant role in boosting human health and well-being (Shellock, 2020). People of all demographic groups visit the coast to spend quality time with loved ones. The ocean is a uniquely accessible global commons which connects us all.

Fisheries provide essential nutrition for three billion people and are just one sector of the ocean economy. Artisanal or small-scale fisherfolk comprise the great majority of the world's fishers. Over 90% (FAO, 2012) live in developing countries and are among society's poorest.

While there is significant dependency on the ocean economy, ocean pollution, climate change, exposure to extreme weather and socio-economic and cultural changes are all anticipated to impinge on human physical and mental health and quality of life. These impacts risk further impoverishing the world's poorest, through loss of livelihoods and economic opportunities.

We are already experiencing more frequent devastating storms; globally significant coral reef loss; migrations of marine species from their habitual ranges; increased toxic blooms and tides and unprecedented levels of plastic pollution. The proportion of global fish stocks being exploited within biologically sustainable limits dropped to 65.8% in 2017, down from 90% in 1990 (FAO, 2020). Small-scale fisheries are further hampered by lack of market access and limited pricing power.

Furthermore, a significant proportion of people in small island developing states (SIDS) live on land with an elevation of 5 m or lower, and face displacement as sea level rises. Global tourism, a mainstay for many island destinations, has all but collapsed amid the pandemic.

The well-being of the ocean and of people are inextricably linked in multiple ways. Both are under serious threat, nowhere more so than in small island states and least developed "large ocean nations".

10.2   THE BLUE ECONOMY

The nascent blue economy concept sets out a bold new vision for more harmonious human interactions with our ocean which is fundamentally different to our current trajectory. The blue economy concept seeks to promote economic growth, social inclusion, and the preservation or improvement of livelihoods while at the same time ensuring environmental sustainability of the ocean and coastal areas (UNDESA, 2017). Furthermore, it recognizes that such a bold new vision must fully anticipate and incorporate the impacts of climate change on marine and coastal ecosystems, requiring a dynamic systems approach.

It is possible to create new sources of value from the ocean in various ways. Nature-based solutions which restore ecosystems such as mangrove forests create value through providing compatible livelihood opportunities such as small-scale fishing and beekeeping, while reducing flood risk to coastal infrastructure and sequestering carbon above and below ground. Greater value can be generated from existing ocean sectors such as fishing, by adopting sustainable fishing methods, turning fish waste into alternative products and adding value post-harvest. New sectors can also be developed, such as offshore renewable energy and marine biotechnology for food and pharmaceuticals. Each country will determine its blue economy strategy according to its unique natural, cultural and economic circumstances.

Realizing the potential of the blue economy and a sustainable quality of life requires multilateral cooperation, innovative approaches, extensive stakeholder engagement and knowledge transfer if it is to live up to its promise of achieving lasting benefits for society and the environment. UN Sustainable Development Goal 14 Life Below Water calls for the development of research capacity to allow SIDS and least developed countries (LDCs) to benefit from marine biodiversity. SDG14 has an influence on almost all other SDGs.

Nowhere is quality of life more deeply interconnected with the ocean than in island nations. The ocean is the lifeblood of islanders, providing social and psychological well-being and economic wealth, typically through tourism and fisheries. People in these countries are already feeling the effects of climate change on their quality of life, such as surrendering land to the sea, coral bleaching and mortality across the tropics with associated loss of marine life, and devastating storm surges. So perhaps it comes as no surprise that these countries have been leaders in championing a blue economy approach which features social equity and environmental sustainability as core tenets (Bennett et al., 2019).

Among these island nations, Seychelles stands out as a beacon of innovation in the blue economy sphere. As blue economy pioneer and former Seychelles President James Michel notes, "the blue economy is about intelligent growth, social inclusion and empowerment in a world that is increasingly challenged by a changing climate" (Michel, 2016).

10.3 SEYCHELLES: A CASE STUDY OF BLUE ECONOMY INNOVATION

> The Seychelles experiment is one of crucial interest to the entire development community, especially in light of climate change agreement negotiations and the guiding principles of the Sustainable Development Goals that will lead us to 2030.
>
> *– Jose Graziano de Silva, Director-General, Food and Agriculture Organisation of the United Nations (Michel, 2016)*

Home to around 94,500 people, the Republic of Seychelles has the smallest population in Africa. It comprises 115 islands scattered across the Western Indian Ocean and with just 452 km² of landmass and an exclusive economic zone encompassing 1,374,000 km² of ocean, the country is effectively 99.97% ocean. This makes it the 24th largest ocean nation in the world. Seychelles' unique island and marine environments are of global significance for biodiversity.

Until 2020, tourism has been the prime source of growth in Seychelles and the fisheries sector is the second key pillar of the national economy, in terms of both foreign exchange earnings and employment. Both of these sectors rely on a healthy ocean. In common with many island nations, Seychelles depends heavily upon imported food, energy and other goods. Diversifying its economy to build personal quality of life and national resilience is essential.

10.3.1 FROM CONCEPT TO REALITY

A wave of innovations is propelling the blue economy concept forward to create a new inclusive and prosperous future for the Seychellois. Some of these are highlighted below.

10.3.1.1 Creating a Long-Term Vision

Seychelles' leaders have consistently advocated for ambitious marine protection to underpin national blue prosperity, in the belief that by acting now to protect their environment, they can be sure they are protecting their people and their livelihoods against an uncertain future. Perhaps this consistent sharing of a hopeful long-term vision can unite generations, providing direction and inspiration, allowing Seychellois society to feel positive and purposeful despite an uncertain future.

According to the World Leadership Alliance Club de Madrid, which aims to strengthen leaders for the development and well-being of citizens, "James Michel has propelled Seychelles on the international scene as an advocate of the cause of small island developing states, the preservation of the environment and the blue economy".

His long-term vision for blue prosperity where people thrive in harmony with the ocean represents an unusual break from short-termism which dominates political agendas and fails to incorporate the needs and rights of future generations to a good quality of life. It embraces the need to set a long-term national agenda in tune with the ocean.

10.3.1.2 Innovation in Finance and Governance

In partnership with various organizations, the Seychelles Government has pioneered some world firsts.

The world's first Debt for Nature Swap for the ocean was announced in 2015 and allowed Seychelles the opportunity to restructure part of its national debt in exchange for committing to protect 30% of its national waters as part of a comprehensive marine spatial plan. A portion of Seychelles' debt repayments now fund innovative marine protection and climate adaptation projects, through the Seychelles Conservation and Climate Adaptation Trust, SeyCCAT.

Building on its success in Seychelles, the Nature Conservancy aims to dramatically scale-up ocean conservation around the world in up to 20 countries over the next 5 years, following this oceanic debt for nature swap model. This initiative aims to protect up to 4 million km² of the world's most critical ocean habitats.

As part of the debt swap agreement, Seychelles is implementing the Western Indian Ocean's first comprehensive large-scale marine spatial plan, and the second largest in the world. IOC-UNESCO defines marine spatial planning as a process of analysing and allocating three-dimensional marine spaces to specific uses, to achieve ecological, economic, and social objectives that are usually specified through a political process. The Seychelles marine spatial plan is the first designed to address climate change and sustain a national blue economy.

In a feat of data fusion, the Seychelles marine spatial plan combines hundreds of data layers which can inform decision-making on optimal and sustainable use of the Seychelles extensive sea space of 1.3 m km². The guiding principles of the plan recognize that "transparency, inclusivity and participation are cornerstones of the engagement, consultation and communication with stakeholders and civil society".

> This effort will help the people of Seychelles protect their ocean for future generations, and will serve as a model for future marine conservation projects worldwide.
>
> – *Leonardo DiCaprio, February 2018*

SeyCCAT was created as an independent organization via a partnership between The Nature Conservancy and the Seychelles Government, as part of the Debt for Nature Swap. SeyCCAT's mission is to strategically invest in ocean stakeholders to generate new learning, bold action and sustainable blue prosperity in Seychelles. It manages the proceeds of the debt swap and funds from the issuance of the world's first sovereign blue bond by the Seychelles Government in 2018, supported by the World Bank and the Global Environment Facility.

The SeyCCAT model is a not-for-profit one, offering grants to local organizations selected via a rigorous and competitive process and looking to conservation organizations and academic partners to ensure that its funds are deployed with impact. The unique strength of SeyCCAT is its ability to operate autonomously, free from political influence with enormous flexibility in the kind of support it can provide to the local people, consistent with its clear mandate to uphold conservation and sustainable use of the Seychelles' marine environment.

The Seychelles' blue economy roadmap was created in 2018 and guides national priorities to 2030. Focus areas proposed for investment are food security and wellbeing; better local food production and markets; access to high-quality education, professional training and employment opportunities; investment in innovation and small and medium enterprises and fostering entrepreneurship.

10.3.1.3   International Diplomacy and Collaboration

Seychelles' leaders have been active proponents of the blue economy internationally and could not have achieved the successes above without ongoing collaboration with partners across all sectors.

Seychelles has a firm tradition of being friends to all. Its "creole diplomacy" (Bueger and Wivel, 2018) has been characterized as demonstrating an openness to and appreciation of difference, holistic thinking and pragmatic problem-solving. It seems people in Seychelles are open to working with others to get things done.

James Michel observes

we need governments and international agencies that can remove bureaucratic barriers and open the gates to new ideas. They must be prepared to take risks and to look beyond their own limited periods of office. Transforming the ocean is a long-term project, on a scale far larger than anything else in human history. It will need a level of cooperation between nations that has never happened before

– (Michel, 2016).

10.3.1.4   A People-Centred Agenda

Let people drive the blue economy agenda and come up with their solutions to the challenges. Grassroots or bottom-up approaches have been far more effective. Here again, we can point to the efforts made to ensuring access to finance.

– Angelique Pouponneau, CEO, SeyCCAT

There can be no blue economy without people from all walks of life making it happen. SeyCCAT has adopted an "inclusive by design" approach since Angelique Pouponneau took the helm in 2018, proactively addressing accessibility by seeking out input from communities, mobilizing existing organizations to help build core skills, and adopting a "come as you are" attitude. Broadening the base of grantees enables SeyCCAT's benefits to reach further and include some of the poorest communities in Seychelles.

SeyCCAT is still a young organization and its first two leaders have quickly developed strong foundations, frameworks and partnerships to enable the organization to deliver ambitious and robust marine science and conservation goals. When Angelique joined SeyCCAT, her focus on widening accessibility to SeyCCAT's funds saw the organization treble its applications with requests amounting to US$ 2.5 million for a US$ 1 million fund in 2019. This was achieved through capacity building, delivered free of charge by partners; permitting grant applications in Creole; and significant direct engagement with fishing communities. For the first time, applicants were not mostly NGOs with significant experience in applying for grants, but new local actors taking a lead.

10.3.2   A QUALITY OF LIFE PERSPECTIVE: METHODOLOGY

As proponents of approaches to the blue economy which are not only environmentally sustainable but also socially inclusive and equitable, our study set out to understand how initiatives supported by SeyCCAT to further the blue economy might improve quality of life in Seychelles, at both an individual and national level.

To date, the SeyCCAT Board has approved 34 projects which seek to advance its strategic objectives. These objectives include improving governance, sustainability, value and market options of the fisheries sector; developing social resilience plans to adapt to the effects of climate change; and trialling and nurturing business models to secure the sustainable development of Seychelles' blue economy.

10.3.2.1   Respondents

From these projects, we selected ten with a direct socio-economic focus rather than a solely environmental one and conducted ten interviews, nine with the project lead and one with a project participant. Table 10.1 sets out the aims of the selected projects

TABLE 10.1
Project Participants

	Project Title	Project Aims
1	Seaweeds: a hidden resource – a recycling project	• To *reduce by an* estimated 15% the amount of seaweed on the beaches on Mahé, especially the east coast. • To produce at least 150kg of high-quality compost for sale to home gardeners. • To increase the income of disadvantaged and vulnerable women and girls engaged in this activity by at least 70%, from minimum wage to earnings of up to SR12,000 monthly.

2	Nou lanmer ble: Lannen 2020	• To produce an underwater documentary series (of ten episodes) in Creole with English subtitles, to inform the public on:
		  o marine species such as sharks, turtles, corals, invertebrates, fish and other species under threat;
		  o climate change;
		  o threats to the ocean and how human activities affects the ocean (e.g., unsustainable fishing practices, plastics pollution, marine conservation).

3	Marine scholarship programme	• To develop a marine science and conservation programme for young adults in Seychelles. • To deliver two 9-month training programmes with at least six participants aged between 18 and 25, including placements with partner organizations. • To build local capacity in the Blue Economy sector by providing educational and vocational opportunities to young adult Seychellois. • To increase access to marine and Blue Economy employment opportunities for young adult Seychellois.
4	Piloting voluntary fisheries zone closure on Praslin island for the benefit of the marine environment and fisherfolks	• To pilot a voluntary fishing zone closure scheme on the island of Praslin, to show the government and the general public that marine areas managed locally by the community can be as effective as those managed by government and NGOs. • To ensure that fishermen have sufficient fish to catch inside the bay during the South East Trade winds when it is difficult to fish beyond the reefs. This should reduce their accident risk and contribute towards their livelihoods.
5	Improving the socio-economic knowledge of the Seychelles Artisanal Fishery	• To develop a methodological guide to produce socio-economic indicators. These will be used to integrate key socio-economic information into policy decisions to promote more effective management of fisheries.
6	Blue economy entrepreneurs – creating smart, sustainable and shared prosperity through entrepreneurship ecosystem assessment and training	To map the entrepreneurship ecosystem within the blue economy and identify the main challenges and potential for growth/improvement. In addition: • Assess the current attitudes of stakeholders towards the blue economy and new 'blue' business models. • Nurture viable 'blue' business models from a 3-day start-up entrepreneurship training workshop that look to integrate environmental and economic sustainability for eventual trial, with the support of local partners. • Identify gaps within the system that hinder entrepreneurial innovation and development in the blue economy. • Share findings with key stakeholders to assist with policy planning matters.

(Continued)

TABLE 10.1 (*Continued*)
Project Participants

Project Title	Project Aims
7 TGMI blue economy accelerator programme	• To provide a platform for aspiring entrepreneurs to access hands-on enterprise skills capacity building to take their ideas to market. • To promote innovation in the fisheries sector in Seychelles. • To accelerate struggling entrepreneurs in achieving investment-readiness status. • To operationalize the blue economy concepts of Seychelles into tangible businesses.
8 Citizen's Guide to Climate Change	• To print and disseminate 1,000 copies of the climate change citizen's guide in both English and Creole. • To support climate change workshops and other climate change education activities.
9 Entrepreneurship development in the blue economy sector through capacity building for MSMEs and ESA staff	• To create a capacity building programme for new enterprises wishing to venture into the blue economy sector.
10 Explore the route to market for seafood from local fishermen	• To create an app to gather online data on fish consumption, sustainability and the route to market.

surveyed. We did not interview the intended beneficiaries of these projects, such as fishermen, women harvesting seaweed and individuals who took part in training and education initiatives and we cannot make any conclusive statements about their quality of life. Although the number of surveys we conducted was small, our choice of projects represents a balanced cross-section of current SeyCCAT projects underway with a clear socio-economic theme and represents almost a third of SeyCCAT-funded projects to date.

10.3.2.2 Materials

Our survey posed around a dozen open questions on factors perceived to affect quality of life and relating to the aims of each project. For example, we asked respondents how they would define quality of life; what factors they considered contribute to a good quality of life; how they thought the ocean affects quality of life; and how their projects could affect quality of life for those involved and more widely.

Each questionnaire was allowed to evolve in the context of the project and in response to the answers given. No pre-existing theory was supposed; our goal was to conduct an open enquiry and allow the responses to suggest themes which can inform future practice and policy both for SeyCCAT and for programmes with similar socio-economic goals.

10.3.2.3 Procedure

Participants were given the option of being interviewed in person, over a telephone call, or submitting written responses. Equal numbers of respondents chose to submit written responses and be interviewed via a telephone call. None of the interviews were conducted in person. The interviews and written responses were all completed over the months of May and June 2020, which coincided with a period of COVID-19 lockdown in Seychelles.

We also asked the former President James Michel how quality of life for current and future generations of Seychellois will be affected by the national blue economy plans, what his vision of success is and the key steps towards achieving this.

10.3.2.4 Analysis

We used a conventional content analysis approach (Hsieh and Shannon, 2005) to allow insights from the survey to emerge. Codes and categories were derived from the data, based on condensing the survey responses into short sections of meaning, grouping similar meaning sections into short-hand codes, and identifying categories which reflected overall trends across these codes. We also noted our initial intuitive responses and compared these with the detailed analyses to form a rounded picture of the survey results.

Simple templates (Erlingsson and Brysiewicz, 2017) for performing this qualitative analysis were adopted. Table 10.2 shows examples of how we assigned codes to condensed units of meaning from the surveys and grouped like codes into categories. For instance, empowerment and inclusion were each either explicitly mentioned or alluded to by at least a third of our respondents, without either of these aspects having been referred to in our survey questions.

Because the survey responses were generally quite expansive and converged in various ways, we could describe qualitative recurring themes from the categories arising.

To counter any bias in interpreting the survey responses, two of the authors conducted a separate analysis as described. We compared the resulting codes, categories and themes to identify points of similarity and difference.

10.3.2.5 Findings

One of the striking results from the survey was that all respondents mention sustainability, without it being explicitly mentioned in our questions, suggesting that they are well aware of the central role of sustainability in shaping the blue economy.

Interconnectedness is described, between ocean ecosystems, the land, climate change and people. Respondents see the systemic way these affect one another, the evolving blue economy and their quality of life.

Almost all respondents view quality of life as having a number of dimensions over and above having essential needs met, including mental well-being, balance, freedom, culture, connection with the environment and hope. Access to opportunities, education, gaining skills, learning, collaboration, benefitting others, acting sustainably, taking climate action and contributing to the country are all mentioned within the survey responses as improving a person's quality of life. Experiencing a good quality of life is about feeling "bien dans sa peau".

TABLE 10.2

Example Codes and Categories Derived from Condensed Survey Extracts

		Categories		
Condensed Meaning Unit	Code	Quality of life	Empowerment	Inclusion/ Ownership
How well we manage our time and balance career, family, hobbies, and spiritual life	Balance	Balance		
Quality of life can include happiness, community, cultural and environmental connection, and freedom.	Freedom	Freedom		
	Happiness	Happiness		
	Culture	Culture		
	environmental connection	environmental connection		
Innovative blue growth needs empowered young people with opportunities, skills and motivation	empowerment		empowerment	
Fishermen becoming more empowered	empowerment		empowerment	
Women have a sense of ownership of the project	Ownership			Ownership
Blue economy to be Seychellois owned	Ownership			Ownership
So it feels like we are part of it	Inclusion			Inclusion
By including people, you have more compliance because you have less conflict	Inclusion			Inclusion

The ocean clearly plays a prominent role in the lives of Seychellois, affecting them both positively, such as offering a place to relax and escape the chaos of work and life, and negatively, such as witnessing increasing coral bleaching. The ocean is seen as a source of opportunity, livelihoods, education and spiritual well-being and as exerting a significant influence on people's living standards.

Collaboration is mentioned by several of the respondents and partnerships are integral to all.

All of the respondents foresee continuance and evolution of their initiatives, and a wider, cumulative impact. Contributing to others and to the country and improving lives are mentioned by several. There is an ambitious feel to many.

Finally, several of the projects are clearly a passion for those leading them. In these responses, the urge to make a positive change is palpable.

Our analysis revealed the following themes.

10.3.2.5.1 Local Empowerment

Greater empowerment of local communities and particular constituencies, such as fishermen, young people and women, is seen as important.

One project led by a fishermen's association to pilot local management of fisheries is seeking to show the Seychelles Government that proactive community action can be effective and impactful, perhaps suggesting that this may be preferable to waiting for regulation. A recent press report notes the success of this initiative, with an increase in the size and number of fish caught following a voluntary closure period initiated by the fishers' association. This greater size and abundance of fish is allowing the fishermen to trap fish in calmer waters of the bay during the southeast monsoon, rather than having to venture further out to sea, making life easier and safer for the fishermen concerned. These press reports note that a nearby fisher community have watched developments with interest and may wish to implement the same process in a bay on their island.

The project lead views this gentleman's agreement as very symbolic, demonstrating a willingness on the part of the fishermen to step forward as a community and try co-management of fisheries with the Government, rather than await a top-down approach.

In addition to making the fishers' livelihoods safer and more predictable, the project has developed greater camaraderie between the fishers, offered them a degree of learning and given them the confidence that comes with taking collective responsibility for natural resources, and implementing a successful pilot project that is now attracting others. It has undoubtedly enhanced their quality of life in a number of important ways.

The artisanal fisheries sector accounts for over 95% of the catch in Seychelles waters (Le Manach et al., 2015) and is of paramount importance for domestic food security, employment, and cultural heritage (Christ et al., 2020). So how small-scale fisheries are governed has significant implications for the quality of life of these people, some of whom are among the poorest communities in Seychelles.

According to a recent synthesis on small-scale fisheries, governance must be responsive to the needs of small-scale fisherfolk. There is no one size fits all (Jentoft and Chuenpagdee, 2015), but optimal approaches foster collaboration, empowerment, and innovation.

A paper commissioned in 2020 by the High Level Panel for a Sustainable Ocean Economy also suggests nation-states embrace new modes of inclusive governance that are framed by an agreed general set of top-down principles but powered by bottom-up decision-making on resource use (Swilling et al., 2020).

Another project we surveyed aims to inspire and empower young adults and provide them with knowledge, confidence and contacts to move into the marine sector, through a marine scholarship programme. In a 9-month training programme, participants undertake 6 months of onsite theory and practical skills development and 3 months in placements in various blue economy workplaces, gaining exposure and building networks.

Providing education, skills, confidence and a diversity of livelihood options, enabling participants to contribute positively to their community and sustainable use and protection of the ocean, are all expected to improve individual well-being.

10.3.2.5.2 Wider Inclusion

An opinion is expressed by one of our respondents that the blue economy should be locally owned. Wider grassroots Seychellois' inclusion is seen as necessary by several. And greater inclusion is also a key focus of SeyCCAT, evident in recent initiatives to develop this, as noted above.

Bottom-up innovation will play a critical role in powering the blue economy, and greater inclusion is a central pre-condition to enable this.

One of the projects we surveyed is converting an over-abundance of seaweed into a productive composting enterprise undertaken by women. The project's proponent hopes that the women involved feel a sense of pride and achievement and a spiritual connection with their environment and the nation, in addition to the financial and social benefits of this "cleaning with purpose". He wishes to give these women enough confidence to talk about their experiences on national radio or television and become role models for others. Ultimately, his aim is for these women to manage the business themselves and employ others.

Another project aims to give visibility to fishermen and those on whom they depend, and in doing so bridge a gap between the Government and fisheries stakeholders to develop a people-centred approach to fisheries management, rather than a traditional fish stock approach.

It seeks to build more constructive dialogue between a large number of diverse stakeholders by conducting an extensive survey to identify common objectives and points of difference, from which to develop shared understanding and dialogue. The project lead expresses the need to gain a deep understanding of the concerns and needs of the most vulnerable, to "be in their skin". Together with others among our survey respondents, she sees fishermen as having a tough job and needing support. Undertaking this project has improved her appreciation and respect for those who contribute to something culturally important, eating a freshly caught local fish.

Representing the views of fishermen and those closely linked to them in an evidence-based analysis is seen as lending weight to decision-making. Wider inclusion in consultation processes is believed to reduce conflict and increase compliance, and policy-makers should recognize the value of developing a greater understanding of the needs of those already working in the blue economy sector.

These projects exemplify how making others visible and giving them a voice brings recognition and respect, shared understanding and a fairness to proceedings, improving quality of life for those concerned.

10.3.2.5.3 Sustainability Is about Adding Value over the Long-Term

Our survey responses as a whole exhibited a clear understanding that the blue economy is about sustainability. Systems thinking was apparent in the ways that people describe linkages between a healthy ocean and healthy people, life on land, the impacts of climate change and the blue economy. This holistic way of thinking is a key asset in grasping this new economic paradigm. Sustainability transformations call for cross-sectoral thinking and approaches, according to a paper prepared for the HLP (Swilling et al., 2020).

The projects we surveyed have typically evolved from earlier programmes and their leaders foresee continuance and evolution of these initiatives, and a gradual cumulative impact. Their SeyCCAT-funded opportunity is part of a longer journey. Contributing to others and to the country and improving lives are mentioned by several respondents. There is an ambitious and dynamic feel to them all.

Fisheries are seen as prominent in the blue economy, which is unsurprising given this is a key pillar of the Seychelles economy, along with tourism. The

SeyCCAT-supported fisheries projects we surveyed seek not to fish bigger and better, but to add value or provide supporting services while maintaining sustainability; or to improve the lives of fishermen through empowering and supporting them and making fishing a safer undertaking.

The circular economy as part of the blue economy is mentioned by one respondent. The blue economy is evolving and offers opportunities as yet unidentified, according to another. Being open to how the blue economy vision is translated by society into reality is key.

Benefitting others, making a national contribution and contributing to sustainable ocean use are all mentioned as bringing greater individual well-being.

10.3.2.5.4   Foster an Entrepreneurial Ecosystem

Developing entrepreneurs and a resilient entrepreneurial ecosystem are seen by some of our survey respondents as important to bring dynamism, wealth and adaptability to the blue economy journey.

Among the projects we surveyed were two with a focus on developing entrepreneurs. The first of these is providing training and capacity development for blue economy entrepreneurs, and the second is providing a blue accelerator programme. Both are catalytic programmes with a longer-term multiplier effect.

Overcoming the culture barrier to entrepreneurialism is seen as a real challenge, according to one respondent. She notes that it is not in people's culture in Seychelles to be entrepreneurial unless a family member is a business entrepreneur. A further hurdle is that people don't know how the blue economy concept translates into activities that can generate income. To change people's attitudes and behaviours requires mentoring and a list of example activities that people can grasp, according to one respondent.

Training and capacity development which include basic ideation and development of business concepts and which promote creativity and innovative thinking are believed to seed resiliency and agility in new entrepreneurs, enabling them to adapt in times of crisis, responds another.

Entrepreneurs need to be introduced to mentors, form partnerships and network with others to succeed. Mentorship is seen as making the difference in galvanizing people from knowledge to action. This should embrace circular economy thinking to ensure that development remains sustainable.

It is suggested by another respondent that the Seychelles Government can support entrepreneurs by increasing the ease of doing business and adopting a sufficiently flexible regulatory framework to embrace yet-to-be-defined activities, so as not to limit innovative ideas. Hosting cross-disciplinary forums, starting incubators and generating networking opportunities for entrepreneurs are other suggestions for how the Government can facilitate and develop the entrepreneurial ecosystem.

Two further projects we surveyed have a distinctly entrepreneurial flavour, the seaweed to compost project outlined previously and a project seeking to support sustainable fishing by providing an app-based route to market for fishermen. The latter project lead believes that entrepreneurs have a duty to help solve problems and improve situations.

Enterprise opportunities are thought by our respondents to improve quality of life in multiple ways; they enhance how people view themselves, broaden horizons, offer choices for decent work, a degree of autonomy and the ability for individuals to adapt and innovate.

10.3.2.5.5   Support Further Education and Workplace Skills

Education and the development of skills are seen as a way for young people and entrepreneurs to diversify their livelihood opportunities and improve their quality of life. More than half the projects we surveyed have an educational or skills development focus.

The marine scholarship programme aims to empower young people by providing skills, motivation and opportunities within the blue economy; it recognizes and bridges a gap between formal education and the workplace, seeking through a blend of theory, practice and placements to open the door to opportunities where young people can express their passion.

This combination of learning-by-doing, theory and practice, when coupled with targeted training for entrepreneurs, mentoring and capacity development, appears to meet an important need for young people making the transition from formal education to finding decent work.

10.3.2.5.6   People Need to See Role Models and Example Activities

Several respondents see the value of serving as or providing role models which can inspire others. In the words of one, people need to see activities that work and that can create an income, to have an incentive to get involved with the blue economy.

To make the blue economy concept more tangible and accessible for those not already involved in it through fishing, for example, concrete examples of activities in new areas or which add value to existing ones are needed.

10.3.2.5.7   Create Widespread Understanding and Awareness
As a Precursor to Behaviour Change

> One has, first, to overcome a common belief that the blue economy will simply be more of the same: more fishing, more shipping, more tourism, etc. There is a qualitative difference that has to be nurtured rather than just explained. Attitudes will change over time but it can never be assumed that everyone has reached the same level of understanding and vision
>
> *– (Michel, 2016).*

James Michel sums up the need to continually share and enhance understanding.

Comments from our survey include that greater general understanding is needed for conservation and sustainability to be effective; gaps in environmental understanding between different age groups are perceived by one respondent to exist, with children more environmentally aware than older generations. Greater understanding is seen by several respondents as leading to caring for the ocean environment and acting mindfully and responsibly.

One project we surveyed is creating an underwater documentary from the Indian Ocean around Seychelles. The project lead hopes to help Seychellois understand the abundance of resources provided by the ocean, to appreciate its magnificence and appreciate why tourists pay so much to visit the country. Her aim is to help Seychellois understand how urgent and serious the problems of the world's ocean are and to show the international community how their actions affect small states like Seychelles, at the forefront of climate and ocean change. Her goal is to foster greater mindfulness, more emotional connection with the ocean and appreciation of how it benefits society, and for people to take better care of nature.

Being informed, learning, feeling connected to the environment, developing a sense of personal responsibility and agency are all cited as improving quality of life in our survey. As the leader of a project on updating the citizens' guide to climate change notes:

> Climate action can make people feel good and make society more pleasant and more cooperative... I think we might make a better society by taking action on climate change.

10.3.2.5.8 Recommendations from Respondents

In addition to the emergent themes above, our survey responses included recommendations for future programmes.

Clear, simple language helps demystify the blue economy. The use of both Creole and English in Seychelles can extend the reach of information and is more comfortable for many people.

Videos capture everyone's attention, with social media and documentaries captivating people more readily than talks or reading materials. Several respondents suggest sharing experiences on national television and radio. These comments suggest that knowledge sharing methods should be culturally attuned to target audiences.

According to one respondent, people like receiving a booklet, which is more likely to be kept than a leaflet.

One participant in our survey questions whether the size and scope of projects funded by SeyCCAT is sufficient to improve quality of life for those involved. Could SeyCCAT fund longer-term projects, particularly given the need to adapt to climate change impacts?

10.3.2.6 Limitations and Further Research

These themes have been identified from a small number of projects funded by SeyCCAT at a particular time in the development of the blue economy in Seychelles. Not all of the findings will be of wider benefit beyond Seychelles, but we hope that some will usefully inform policy initiatives seeking to improve quality of life while delivering sustainable economic models of development.

Research on the quality-of-life experiences of communities in other large ocean nations with differing cultures where the blue economy is under development would provide valuable additional perspectives to inform policy-makers, practitioners and academia.

10.4   LESSONS FOR FUTURE BLUE ECONOMIES

What lessons can we draw from the case of Seychelles in experimenting with the blue economy approach? What are local perspectives on this, as we zoom from the macro view to the micro?

Our survey has demonstrated a positive correlation between the sustained vision of successive Seychelles' leaders for the national blue economy and of some of its citizens in the vanguard of driving this. Both groups seek and envision a people-centred, locally owned agenda, built around sustainable use of ocean resources, powered by a thriving entrepreneurial community, working in partnership across disciplines, sectors and societies and which gives full voice to those already working in the sector and facing increasing challenges. Both constituencies also wish to see Seychelles as a leader in the blue economy sphere. It is clear that partnerships with a holistic viewpoint are driving the Seychelles blue economy.

The view from our respondents is that efforts to support widespread shared understanding and the development of relevant skills, knowledge and networks should be sustained and expanded to increase inclusion and widen access to opportunities even further.

This view is endorsed by a recent study on the African blue economy experience. Inclusion and ownership, together with transparency in the sustainable use of resources, are benefits that accrue from a collaborative approach to managing natural resources. This requires giving local communities a voice by providing them with the necessary education and capacity building to participate effectively (Okafor-Yarwood et al., 2020).

Flexibility and openness from Government is sought by some respondents to consider ideas and opportunities that will arise and to engage in dialogue with communities. Some of our respondents believe and have shown that they have solutions and can provide essential baseline data for evidence-based decision-making without waiting for regulation. A truly inclusive approach to the blue economy means creating an environment which affords fishing communities and others opportunities to directly manage their environment and by doing so influence their own social and cultural well-being.

Accessing this micro view of the blue economy has allowed the authors to appreciate that the blue economy is both a journey and a destination, and that those at the cutting edge have levels of drive and ambition at least equal to the leaders of Seychelles who have navigated the critical first steps of the journey. A key question that arises from this survey for consideration by SeyCCAT and government stakeholders is how to develop the enabling financial and governance frameworks to increase inclusion and support sustainable and informed decision-making, to allow the people of Seychelles to build on these strong foundations and drive their own agendas.

The smallest country in Africa is forging a powerful example of a new sustainable and socially inclusive pathway for development in tune with the natural environment. This is rooted in compassion and a profound respect for the natural world. It invests time to understand the needs of some of the most vulnerable in society and recognizes the benefits of developing an entrepreneurial culture which embraces sustainability and fosters autonomy and resilience.

As individuals, our well-being is improved when we connect with others, keep learning, stay active, take notice of the world around us and give to others. The Seychelles approach to the blue economy gives us an example of a country which is super-connected; keeps learning by inviting innovation to the table; is proactive in world diplomacy; has a long track record of noticing and caring for its beautiful natural surroundings and understands that the best gift to future generations is a well-cared for environment.

A clear long-term vision for inclusive blue prosperity in Seychelles is becoming a mainstream national agenda embraced by many, promoted by successive Seychelles leaders through the consistent adoption and implementation of key mutually reinforcing objectives, principles and messages. This vision resonates deeply with an intuitive understanding exhibited by all our survey respondents that the health of the ocean, of life on land and the well-being of people are deeply intertwined.

This long-term vision for Seychelles encourages and unites its people in imagining better possibilities. In the current climate, perhaps the ultimate gift to individual and national perspectives on a good quality of life is hope for the future.

ACKNOWLEDGEMENTS

The authors would like to extend heartfelt thanks to the interviewees who took time out of their busy schedules to speak with us. Without them, there would be no chapter. The interviewees include Benjamin Vel, Consultant (Seaweed: A Hidden Resource – a Recycling Project); Dillys Pouponneau (Nou Lanmer ble: Lannen 2020); Georgina Beresford, WiseOceans (The Marine Scholarship Programme); Jude Bijoux, Project Advisor Praslin Fishers Association (Piloting voluntary fisheries zone closure on Praslin Island for the benefit of the marine environment and fisher folks); Karine Rassool, PhD Research Student at the University of York (2020)/Senior Economist – Seychelles Fishing Authority (Improving the socio-economic knowledge of the Seychelles Artisanal Fishery); Malshini Senaratne, Director, Eco-Sol (Blue Economy Entrepreneurs); Dr Michele P. Martin (Citizens' Guide to Climate Change); Marie-Celine Zialor, mceline@tgmi.edu. sc (TGMI Blue Accelerator Programme); R. Pillay (Entrepreneurship Development in the Blue Economy Sector through capacity building for MSME's & ESA Staff); and Mr. Shahiid Melanie (Explore the route to market for seafood from local fishermen).

The authors also wish to thank Dr. Emma Holdsworth, Senior Lecturer in Forensic Psychology at Coventry University, UK, and Jonathan Turner and Andy Hamflett, Directors at NLA International for guidance on interpreting results and chapter content.

Finally, the authors wish to acknowledge the contribution of the former President of Seychelles, James Michel, for his personal views on the potential of the Blue Economy.

REFERENCES

Bennett, N.J., Cisneros-Montemayor, A.M., Blythe, J. et al. 2019. "Towards a sustainable and equitable blue economy." *Nature Sustainability 2:* 991–993.
Bueger C., Wivel A. 2018. "How do small island states maximize influence? Creole diplomacy and the smart state foreign policy of the Seychelles." *Journal of the Indian Ocean Region 14:* 170–188.

Christ H.J., White R., Hood, L. et al. 2020. "A baseline for the blue economy: Catch and effort history in the republic of Seychelles' domestic fisheries." *Frontiers in Marine Science* 7: 269.

Erlingsson C., Brysiewicz P. 2017. "A hands-on guide to doing content analysis." *African Journal of Emergency Medicine 7*: 93–99.

FAO, UN. 2012. *Voluntary Guidelines for Securing Sustainable Small-Scale Fisheries in the Context of Food Security and Poverty Eradication – Facts and Figures.* Online, UN FAO.

FAO, UN. 2020. *The State of World Fisheries and Aquaculture 2020.* Online.

Hoegh-Guldberg, O. et al. 2015. *Reviving the Ocean Economy: The Case for Action.* Online, Geneva. Gland, Switzerland: WWF International.

Hsieh H., Shannon S.E. 2005. "Three approaches to qualitative content analysis." *Qualitative Health Research 15*: 1277–1288.

Jentoft S., Chuenpagdee R. 2015. "Enhancing the governability of small-scale fisheries through interactive governance." In *Interactive Governance for Small-Scale Fisheries.* Eds. R. Chuenpagdee, S. Jentoft, 727–747. Switzerland: Springer.

Le Manach F., Bach P., Boistol, L., Robinson, J., Pauly, D. 2015. *Artisanal Fisheries in the World's Second Largest Tuna Fishing Ground — Reconstruction of the Seychelles' Marine Fisheries Catch, 1950–2010.* Online.

Michel, J.A. 2016. *Rethinking the Oceans, Towards the Blue Economy.* Paragon House. St Paul, MN.

Okafor-Yarwood I., Kadagi N.I., Miranda, N.A. et al. 2020. "The blue economy–cultural livelihood–ecosystem conservation triangle: The African experience." *Frontiers in Marine Science 7*: 586.

Shellock R., 2020. *The Well-Being and Human Health Benefits of Exposure to the Marine and Coastal Environment.* Online, London: UK Government.

Swilling, M., Ruckelshaus M., Brodie Rudolph, T., et al. 2020. *The Ocean Transition: What to Learn from System Transitions.* Online, World Resources Institute. Washington D.C.

UNDESA, World Bank. 2017. *The Potential of the Blue Economy: Increasing Long-Term Benefits of the Sustainable Use of Marine Resources for Small Island Developing States and Coastal Least Developed Countries.* Online.

11 Spatial Assessment of Quality of Life Using Composite Index
A Case of Kerala, India

Fathima Zehba M. P., Mohammed Firoz C., and Nithin Babu
NIT Calicut

CONTENTS

11.1   INTRODUCTION

The concept of Quality of life (QOL), being viewed differently in different communities, is multidimensional in nature. It includes the 'full range of factors that influence what people value in living, beyond the purely material aspects' (EUROSTAT 2015; Young 2008). When it comes to measuring QOL, a number of approaches have been developed by researchers around the world. While GDP measures the economic progress of a country, it may not reflect the actual realities about the living conditions of its people. Such conventional measures succeed in assessing the financial health of a nation, which many consider as a representation of its success, in terms of providing its citizen with a higher standard of life. But in reality, they indicate the ability of a community to progress financially and its resilience to fluctuations in the economy. Economist Simon Kuznets once stated that 'the welfare of a nation can scarcely be inferred from a measurement of national income' (*Handbook of Community Well-Being Research* 2017). After thorough observation of national economies researchers proved that increase in financial ability of a community creates happiness only up to a certain threshold beyond which income increased but happiness did not, a concept called Easterlin Paradox (Freimann 2016). It is in this context that the need to measure QOL arises as it helps policy makers, planners and the administrators to take the right decisions at the right time. Though there has been a huge increase in the amount of research in QOL assessment, the spatial assessment/geographical aspects of the same are little explored, especially in the Indian sub-continent. Spatial Assessment of QOL helps in understanding what are lacking where, identifying areas requiring special focus and formulating policies and strategies for a balanced and sustainable development (Serag et al. 2013).Outlier regions and clusters that have low QOL can be identified through spatial analysis (Askam and Corrado 2012). Therefore, this research discusses a study conducted to assess spatially, the QOL in the settlements of Kerala, a southern state of India. The study was done by proper conceptualization of QOL, formulating a composite index for the study area and spatially analysing the generated index using GIS to identify various zones of higher and lower well-being.

In spite of being part of a developing nation, Kerala shows a different trend of continuous urban agglomeration pattern with high Human Development Index (HDI), female literacy rate and life expectancy, sex ratio and low infant mortality rate when compared to that of other states in India (Franke and Chasin 1989; Paul 2017; Firoz 2014; Krishnan and Firoz 2020; Tharakan 2006, 2008; Kallingal and Firoz 2017). This model of socio-economic development despite poor economic progress has gained a name called the 'Kerala model of Development' (Kallingal 2017; Kallingal and Firoz 2017; Tharakan 2006, 2008; George and Kumar 1997; Franke and Chasin 1989; CDS-UN 1975).

However, certain researchers like (Kallingal 2017; Kallingal and Firoz 2017; Naidu and Nair 2007; Subrahmanian and Prasad 2008; Tharakan 2008; Zachariah and Rajan 2014) are of the opinion that the 'Kerala model of development' is not an inclusive socio economic development and several disparities do exist between the regions and socio economic groups. An understanding of these disparities and the QOL of inhabitants where they live (spatially) becomes a prime necessity in formulating development policies for a comprehensive inclusive development of the state. Measuring QOL of a region using an already established scale becomes difficult for a developing country because of the wide socio-cultural differences between regions, data constrains, etc. Hence, it was necessary to construct a unique scale to measure the QOL of a region using indicators relevant to the social and demographic conditions of the area of concern.

11.2 RESEARCH BACKGROUND

The literature was reviewed under three heads, namely concepts related to QOL, selection of suitable variables of QOL from a holistic literature and selecting appropriate tools and techniques for a composite index construction.

11.2.1 INDICES FOR MEASUREMENT OF QUALITY OF LIFE

Measurement of QOL hasn't been a novel idea in many countries around the world. It is a usual or necessary practice in many developed countries and numerous researches are being conducted on this. In the United States, there are around 200 indicator systems for measuring progress in various policy areas (Young 2008). Most of these indicators use objective parameters which are quantifiable and universally recognized. These systems of measurement are found to be widely accepted and highly useful (Damen 2014). Several indices have been developed around the world for this purpose. Physical Quality of Life Index (1970), developed by sociologist Morris MD, is one among the first attempts to calculate QOL based on basic literacy, infant mortality and life expectancy (Morris 1980). World Happiness Report, developed by United Nations combines both objective and subjective measures to rank countries by happiness. This uses survey results from the Gallup poll, GDP per capita, healthy life expectancy, perceived freedom, etc. Happy Planet Index (2006) measures country's ecological footprint as an indicator in addition to standard determinants of well-being. City Livability Index 2017 (Government of India) assesses cities on a comprehensive set of 79 parameters including availability of roads, education, healthcare and employment opportunities (MUD-GOI, n.d.). OECD Better Life Index provides a detailed set of comparable well-being indicators under 11 domains, used for measuring and ranking countries for its QOL (OECD, n.d.). The Social Progress Index was based on three main dimensions, namely Basic human needs, Foundations of well-being and Opportunity. Basic human needs denote people's capacity to survive which include sanitation, clean water, shelter, etc. Foundations of well-being indicates factors which acts as stepping stones for the people to build a better life like basic education, information access and communication facilities. Opportunity encapsulates factors ensuring equality within a community such as social inclusion of backward communities, access to advanced educational facilities and institutions (Porter, Stern, and Green 2017). The 12 indicators and 51 variables were selected based on their focus, credibility of sources, expert feedback and data availability. PCA was used for aggregation (Kapoor, Kapoor, and Krylova 2016). The Canadian Index of Well-being (CIW) focused on both subjective and objective dimensions of well-being. Data were collected through primary and secondary surveys and given equal weightage stating Laplace's Principle of Non-sufficient reason. Here, simple mean aggregation was used to compute the index (Michalos et al. 2011).

Developing Countries like India are in a state of high demand for a QOL assessing method, owing to its urbanization trends. The indices from developed countries are not directly applicable here due to its contextual differences and data availability. Although the methodology for calculating a QOL index shows similarities in most of the studies (Delsante 2016), context-specific variables may be needed to carry out an assessment in developing countries.

11.2.2 Variables of Quality of Life

Variables of QOL are the foundations of any QOL measurement tool as they define the characteristics of the index. Variables should be representative, simple, valid and policy relevant for a better result and interpretation (OECD, n.d.). Selection of variables of QOL was not given required emphasis in most of the studies, rather followed the available literature. As the major objective of the study was to explore the spatial correlations of QOL and its determinants, the tedious process of finalization of domains and variables was shortened by adopting the same from conceptually similar indices like Social Progress Index, Genuine Progress Indicator, Life Situation Index, Australian Unity Well-being Index, HDI, Happy Planet Index, CIW, OECD Better Life Index, Sustainable Society Index, Gross National Happiness, as well as several other researches in the domain. This initial list of 79 variables were further subjected to alterations according to the contextual significance, ease of understanding, reliability, validity, periodical updating and deletion of overlapping nature of variables. The same was subjected to an expert opinion survey and focus group discussions thus grouping them into three larger domains, seven subdomains and 23 variables. The process of variable selection is shown in the flowchart below (Figure 11.1) and the final list of variables selected for the study is shown in Table 11.1.

11.2.3 Assessment Methods and Techniques

QOL assessment methods were largely found to be context specific. A common approach is to use an already available international QOL index and apply it to the concerned case. In such a case, it is not always recommended because, a universally acceptable Index cannot capture the local variability of the context of the study. Therefore, for studies that involve higher number of variables (usually more than 10) and with those with variables more related to the context and locality of the study region, it becomes imperative to device a new scale of measurement rather than depending on the universally accepted global indices. Hence, use of multivariate techniques becomes essential, for example, principal component analysis (PCA), factor analysis, Cronbach co-efficient alpha and cluster analysis (OECD 2008) are used in the construction process of a composite index. Other methods include Data Envelopment Analysis, Benefit of the Doubt approach, unobserved components model, Budget allocation Process, Public opinion, etc. (OECD 2008). It can be seen that many of the internationally accepted indices like Weighted Index of Social Progress (WISP), Australian Unity Well-being Index, Gross National Happiness Index, etc. used PCA, while some like Genuine Progress Indicator used arithmetic mean to aggregate the results. Major advantages of using PCA are such that it does

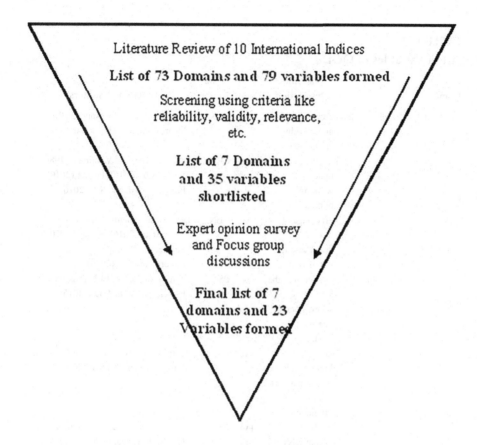

FIGURE 11.1   Variable selection process.

An inverted pyramid showing the process of variable selection saying 'literature review of 10 international indices' at the top and 'final list of 7 domains and 23 Variables formed' at the bottom.

not require large computations, it is adopted by many international Indices and it reduces dimension of data without losing much information (Erik and Marko 2011). Spatial and temporal coverage of the indicators were also considered to cross check its validity and hence PCA is adopted in the study.

In many studies, spatial analysis was carried out with the help of geographical database to analyse the disparities in QOL (Martínez 2009). Homogeneous groups are clustered for discovering patterns of spatial association, concentration areas or hotspots which is called hotspot analysis (Santos and Martins 2014; Ilic 2013). Areal units like wards/divisions/districts (block panchayats in this study) can be scored with respect to the perceptions of citizens (subjective) on the objective indicators through a questionnaire survey (Janagraha 2013). The links between spatial urban forms and QOL can be explored through trend surface maps (Bardhan, Kurisu, and Hanaki 2011). In computing a composite index, the methodological approach used

TABLE 11.1
List of Variables of QOL

Domain		Variable Description	Sign	Data Source	Citations (Authors)
Opportunity	Employment	Percentage of non-working population	−ve	CEN	Scott Stern (2016), Posner (2010), Firoz, Banerji, and Sen (2016)
		Percentage of male non-working population	−ve	CEN	Dutt, Monroe, and Vakamudi (1986), Scott Stern (2016), Posner (2010), Firoz, Banerji, and Sen (2016)
		Percentage of female non-working population	−ve	CEN	Dutt, Monroe, and Vakamudi (1986), Scott Stern (2016), Posner (2010), Saisana, and Philippas (2012), Firoz, Banerji, and Sen (2016)
	Education	Number of higher secondary schools per 1,000 population	+ve	PS	Scott Stern (2016), El-Farouk (2018), Firoz, Banerji, and Sen (2016)
		Number of higher studies institutions per 1,000 population	+ve	PS	Scott Stern, (2016), El-Farouk, (2018, Firoz, Banerji, and Sen (2016)
		Number of anganvadis per 1,000 population	+ve	PS	Firoz, Banerji, and Sen (2016), Zehba (2018), Kallingal (2017)
	Services	Number of banks and cooperative institutions per 1,000 population	+ve	PS	Scott Stern (2016) Firoz, Banerji, and Sen (2016), Kallingal (2017), Zehba (2018)
		Length of road per square kilometer of the study area	+ve	PS	Firoz, Banerji, and Sen (2016), Zehba (2018), Kallingal (2017)
		Number of Post offices per 1,000 population	+ve	PS	Firoz, Banerji, and Sen (2016), Zehba (2018), Kallingal (2017)
		Number of Public distribution shops (ration shops) per 1,000 population	+ve	PS	Firoz, Banerji, and Sen (2016), Kallingal (2017), Zehba (2018)

(Continued)

TABLE 11.1 (*Continued*)
List of Variables of QOL

Domain	Variable Description		Sign	Data Source	Citations (Authors)
Foundations of well-being	Access to basic knowledge	Total percentage of the population that is literate	+ve	CEN	Dutt, Monroe, and Vakamudi (1986), Madu (2010), Maynooth and Martin (2010),(Bogdanov, Meredith, and Efstratoglou 2008), Chi and Ventura (2011), Monasterolo and Coppola (2011), Firoz, Banerji, and Sen (2016)
		Percentage of adult population that is literate	+ve	CEN	Dutt, Monroe, and Vakamudi (1986), Madu (2009), Ogdul, and Getal (2010), Maynooth (2000), (Bogdanov, Meredith, and Efstratoglou 2008), Chi and Ventura (2011), Monasterolo et al. (2010), Firoz, Banerji, and Sen (2016)
		Percentage of female population that is literate	+ve	CEN	Dutt, Monroe, and Vakamudi (1986), Ogdul, and Getal (2010), Maynooth (2000), (Bogdanov, Meredith, and Efstratoglou 2008), Chi and Ventura (2011), Monasterolo et al. (2010), Firoz, Banerji, and Sen (2016)
		Number of schools (LP/UP and HS) per 1,000 population	+ve	PS	Firoz, Banerji, and Sen (2016), Zehba (2018), Kallingal (2017)
		Number of anganvadis per 1,000 population of 0–6 age group	+ve	PS	Firoz, Banerji, and Sen (2016), Zehba (2018), Kallingal (2017)
	Social equality	Percentage of SC/ST households electrified	+ve	PS	Firoz, Banerji, and Sen (2016), Zehba (2018), Kallingal (2017)
		Percentage of SC/ST households with water connection	+ve	PS	Firoz, Banerji, and Sen (2016), Zehba (2018); Kallingal (2017)
		Percentage of SC/ST households with pour or flush latrine connection	+ve	PS	Saisana, and Philippas (2012), Firoz, Banerji, and Sen (2016)

(Continued)

TABLE 11.1 (*Continued*)
List of Variables of QOL

Domain		Variable Description	Sign	Data Source	Citations (Authors)
Basic human needs	Shelter	Percentage of Kutcha houses	−ve	PS	Mehdi (2019), Firoz, Banerji, and Sen (2016)
		Percentage of electrified households	+ve	PS	Mehdi (2019), Firoz, Banerji, and Sen (2016)
	Health and wellness	Hospitals per thousand population	+ve	PS	Posner (2010), Scott Stern (2016), Firoz, Banerji, and Sen (2016), El-Farouk (2018)
		Average number of beds per hospital in the study area	+ve	PS	Posner (2010), El-Farouk (2018), Firoz, Banerji, and Sen (2016)
		Number of doctors available per 1,000 population	+ve	PS	El-Farouk (2018), Scott Stern (2016), Posner (2010), Firoz, Banerji, and Sen (2016)

Note:
Anganvadis – Kindergartens.
+ve Symbol of the variable shows that higher the value is good for QOL.
−ve Symbol of the variable shows that higher the value is bad for QOL.
PS, Panchayath level Statistics, Government of Kerala; CEN, Census of India 2011.

in each research was different. Hence while choosing, the social, demographic and geographical conditions of the context had to be considered.

11.3 MATERIALS AND METHODS

11.3.1 STUDY AREA

The south Indian state of Kerala as a whole was chosen as the study area (refer Figure 11.2) with unit of study as Block Panchayat,[1] as it helps to understand the geographical patterns more precisely. As already mentioned, Kerala is known to have high HDI values, socio demographic indicators and high health scores comparable to developed countries. Also, the settlement pattern follows a rural-urban continuum where rural and urban areas cannot be differentiated (Firoz 2014; Krishnan and Firoz 2020). Despite these facts, there exist disparities in the living conditions of people at the settlement level. Studies on understanding the QOL disparities in Kerala haven't been done and hence become a necessity.

---

[1] The administrative set up of India adopts a decentralized system comprising the country at the apex, followed by the state and the local self-government. Each state is comprised of districts and each district is further divided as urban local bodies named as Municipal Corporations or Municipalities and rural local bodies named as Gram Panchayats.

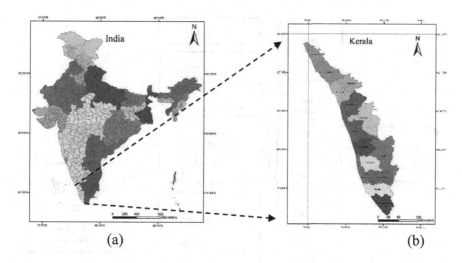

FIGURE 11.2   Location of study area. (Krishnan and Firoz 2020).

Map of the country India from which map of the state of Kerala is projected out and shown bigger.

The study is using 2011 Census data and Panchayat Statistics.

The methodological framework of the study is shown in the flowchart (Figure 11.3). The whole process was done mainly in two steps; construction of a composite index of QOL and spatial analysis. At first, variables were finalized through shortlisting, as explained in the literature review Section 11.2.2. A variable matrix with 151 block panchayats and 23 variables was formed and corresponding data were collected. The major sources of data were secondary such as Census of India and Panchayat Statistics (Babu 2019). The data were then made go through a series of statistical analysis to finally compute a composite index formula for assessing QOL.

11.3.2   CONSTRUCTION OF COMPOSITE INDEX

The variables of QOL had to be weighted and aggregated to get the overall well-being of any study area (Risser et al. 2017; OECD 2008). For this, a composite index of QOL was constructed. Once the variable matrix was finalized, data were collected from secondary sources such as Census of India and Panchayat statistics. As explained in the overall methodology, the composite index construction process involves two stages, namely initial data exploration and multivariate analysis (Erik and Marko 2011; Babu 2019; Kallingal 2017; Zehba 2018).

11.3.2.1   Initial Data Exploration

At first, the data matrix was explored using descriptive statistics such as range, mean, median, standard deviation, variance and coefficient of variation. This was needed to understand the characteristics of the data set (Nardo et al. 2005). This also helped in identifying the missing values for each variable. Since data were collected from

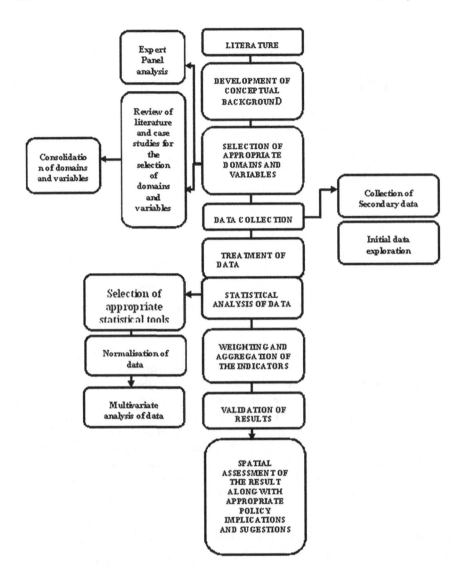

FIGURE 11.3 Methodological framework.

A flowchart of overall methodological framework, starting from 'literature' through 'development of conceptual background', 'selection of appropriate domains and variables', 'data collection', 'treatment of data', 'statistical analysis of data', 'weighting and aggregation of indicators', 'validation of results' and ending at 'spatial assessment of the result along with appropriate policy implications and suggestions'. The box 'selection of appropriate domains and variables' branches out to 'expert panel analysis and 'review of literature and case studies for the selection of domains and variables' and ends at 'consolidation of domains and variables'. The box 'data collection' branches out to 'collection of secondary data and ends at 'initial data exploration'. The box 'statistical analysis of data' branches out to 'selection of appropriate statistical tools' through 'normalization of data' and ends at 'multivariate analysis of data'.

different sources, there were cases of missing values in the data set. As missing values would hinder the robustness and results of any statistical analysis, these had to be properly treated. Out of the many methods for this such as mean-value imputation, K-nearest neighbours (KNN), fuzzy K-means (FKM), singular value decomposition (SVD), Bayesian principal component analysis (bPCA), multiple imputations by chained equations (MICE), etc., Unconditional Mean Imputation in explicit modelling of Single Imputation (Nardo et al. 2005) was chosen in which the missing values were replaced with the mean value of the group of concern. Data set was then checked for any accidental errors and was corrected manually before doing any further analysis. A multivariate analysis was then carried out to evaluate the fit of the data set.

11.3.2.2 Multivariate Analysis

Multivariate analysis is used to analyse the aptness of data set and to check whether different dimensions of the phenomenon are statistically well balanced. Here, it is carried out as an exploratory analysis to investigate the overall structure of the indicators, assess the suitability of the data set and explain the methodological choices in weighting and aggregation (OECD 2008; Becker et al. 2017). As mentioned earlier in Section 11.2.3, PCA was used in this study for the construction of a composite index. This involves stages, namely check for normality, outliers and multicollinearity, normalization and weighting and aggregation (Erik and Marko 2011; Nardo et al. 2005).

11.3.2.2.1 Check for Normality and Outliers

A normality check is done to understand spread of the data and statistical significance of the data set. There are many methods of performing a normality check such as normal probability plot, a quantile–quantile plot (QQ plot) of the standardized data against the standard normal distribution, Histograms, etc. (Nardo et al. 2005). As sample size increases, it is more difficult to check the normality of data set since the parameters are more restrictive. So for large data sets, normality testing is considered less important (Ben-Gal 2005). Hence, in this study, the normality check was not performed due to its irrelevance.

Outliers, if not treated and removed would affect the modelling and statistical analysis of the data set. They will result in model misspecification, biases parameter estimation and incorrect results (Ben-Gal 2005). Outliers could be identified through boxplots, scatter plots, etc. (visualization tools), or Z-score, IQR score, etc. (through mathematical functions). In this study, box plot diagrams were used and were created using SPSS (Erik and Marko 2011; Nardo et al. 2005). Once the outliers were identified, they were eliminated by restricting and capping the data set to 5th and 95th percentile values of the mean value of the data set.

11.3.2.2.2 Check for Multicollinearity

Multicollinearity is a case of multiple regression in which the variables are highly correlated between themselves. If the variables are very low or very high correlated, it could alter the results of PCA (Nardo et al. 2005). In this case, multicollinearity was expected because of the errors in the data collection methods and constraints in the population. Correlation analysis was conducted to check these two phenomena and the variables were merged or deleted accordingly.

11.3.2.2.3   Normalization

Normalization is a scaling technique where a new range of values is found from the existing range of data set to make the data statistically normal. This is done because intervariable comparison would not be possible if data collected were of different units or scales. Different methods like standardization, min-max transformation, etc. are used widely for normalizing data (Nardo et al. 2005). Here, a commonly used Z-score transformation was used for normalizing the data (OECD 2008).

11.3.2.2.4   KMO and Bartlett's Test

'The Kaiser-Meyer-Olkin (KMO) measure of sampling adequacy is a statistic for comparing the magnitudes of the observed correlation coefficients to the magnitudes of the partial correlation coefficients' (OECD 2008). It is suggested that KMO measure should be greater than 0.5 to proceed with PCA (Erik and Marko 2011; Nardo et al. 2005). Also, the Bartlett's significance value should be less than 0.05, for the data set to be statistically significant for further analysis and modelling (Erik and Marko 2011; Nardo et al. 2005). In PCA, for better reliable results, a KMO value greater than 0.8 was ideal (Erik and Marko 2011; Nardo et al. 2005). The KMO measure for this data set was calculated in SPPS software.

11.3.2.2.5   Principal Component Analysis

PCA is a dimension reduction technique with an objective to present a huge data set using a few variables (Erik and Marko 2011). In this, the variance of the observed data is explained through a few linear combinations of the original data, called principal components. Eigenvalues are the variances of the principal components. Several studies suggest that eigenvalue greater than 1 is the criterion for decision of number of principal components (Erik and Marko 2011).

The input for PCA in this study was the transformed and normalized variable matrix. The principal components were extracted using SPSS software. The component loadings for each variable were obtained and were rotated using Varimax rotation to get a better interpretable result (Erik and Marko 2011; Babu 2019; Zehba 2018; Kallingal 2017).

11.3.2.2.6   Weighting and Aggregation

Once the data set was checked for suitability, statistical significance and was transformed accordingly, the variables were given weightage. Weighting is important since every variable is different and has its own percentage of representation of the composite index (Mazziotta and Pareto 2013). PCA was used here for weighting the variables.

Component scores of each areal unit on each principal component were calculated first. 'Component score are calculated by calculating the case's standardized value on each variable, multiplying by the corresponding loading of the variable for the given principal component factor, and summing up these products' (OECD 2008; Becker et al. 2017). Then, weights were assigned for each component as the proportions of explained variances of each corresponding variables in the data set. Using the proportions as weights on the factor scores and summing up these values of each component, an index was calculated for each settlement (Babu 2019; Kallingal 2017; Zehba 2018). A standardized index was further calculated for better interpretation. Final formula used is shown.

$$\text{Composite Quality of Life Index (non-standardised)}$$

$$= \left(\frac{\text{Var1}}{\text{tot var}}\right) \times (\text{factor1 score}) + \left(\frac{\text{var2}}{\text{tot var}}\right) \times (\text{factor2 score}) + \left(\frac{\text{var}_n}{\text{tot var}}\right) \times \left(\text{factor}_n \text{ score}\right)$$

$$\text{Standardised Composite Quality of Life Index of Settlement1}$$

$$= \frac{\text{Index of settlement1} - \text{Min index}}{\text{Max index} - \text{Min index}} \times 100$$

11.3.2.3 Validation of Results

Validation of the results obtained is required to assess the quality of the index and its reliability. In this study, qualitative validation was done by comparing the results with known information about the settlement. Quantitative validation was done by cross-validation technique. This technique evaluates predictive models by partitioning original samples into a training set to train the model and a test set to evaluate it. If the result is valid, then the Pearson coefficient will be the same for both the actual result and the predicted model (Zehba 2018; Kallingal 2017).

11.3.3 SPATIAL ANALYSIS

Spatial analysis of QOL was carried out in the next stage. The pattern was studied by mapping the scores of the settlements spatially with the help of GIS. As any policy directive solutions should focus on where it is needed, classification of settlements was required (Janagraha 2013; Teklay 2012; Bardhan, Kurisu, and Hanaki 2011). It would help in decentralized planning with focused service provisions and development initiatives. The settlements were classified into three groups – high, medium and low QOL settlements (Zehba 2018; Babu 2019). The range of QOL index scores for this was decided by the standard deviation classification method. The characteristics of each group were studied and policy recommendations were suggested for the betterment of QOL (Babu 2019).

11.4 RESULTS

The final list of variables is shown in the following table. Once data matrix was ready, index scores were calculated through the statistical analysis, as explained in the methodology.

11.4.1   Composite Index of Quality of Life

11.4.1.1   Initial Data Exploration Results

Descriptive statistics of the data set revealed that range of some variables was too high. This indicated the presence of outliers in the data set. Variance was also high for some variables, indicating that the data set was not normally distributed. Standard error of mean for most of the variables was closer to zero, indicating that data set represented the given population almost accurately without much statistical corrections.

11.4.1.2   Multivariate Analysis Results

For checking the presence of outliers, as mentioned earlier in section, Q-Q plot diagrams from SPSS was used here. Normality check was irrelevant in this study due to large sample size (as mentioned earlier) and was not performed.

Outliers were identified using box plot analysis in SPSS and the number of outliers for each variable was found out. It was found that the variable 'percentage of kutcha houses' had the highest number of outliers (7). Outliers were eliminated through capping method, as explained earlier.

Correlation matrix of the variables was found using tools in SPSS for checking multicollinearity. Variables with the Pearson correlation value nearest to unity were identified as highly correlated with each other. Eight out of the 23 variables showed high correlation. One among the two highly correlated variables was dropped in order to deal with multicollinearity. The eliminated variables are Male non-workers %, Female non-workers %, adult male literates %, adult female literates % and Anganvadis per 1,000 population %. Other variables were either normally correlated or not significantly correlated. This was considered good because it made the data less redundant. After correlation analysis and correction enough, the number of variables reduced to 18. After normalizing the data set using Z-score, KMO and Bartlett's test was conducted.

11.4.1.2.1   KMO and Bartlett's Test

The data set showed a KMO factor of 0.768 which was greater than 0.5, as shown in Table 11.2 (KMO and Bartlett's test result). Also, the Bartlett's significance value was 0 which was less than 0.05. As mentioned before, these indicators suggested that the data set was statistically significant and PCA can be conducted.

For weighting the variables, PCA was carried out in SPSS to the rotated Z-Score matrix. The next task was to identify number of components that were statistically

TABLE 11.2
KMO and Bartlett's Test Result

KMO and Bartlett's Test

Kaiser–Mayer–Olkin measure of Sampling adequacy		0.768
Bartlett's test of sphericity	Approx. Chi square	1023.170
	Df.	153
	Sig.	0.000

Source: SPSS Output.

TABLE 11.3
Principal Components

PCA: Total Variance Explained

Component	Initial Eigenvalues			Rotation Sums of Squared Loadings		
	Total	% of Variance	Cumulative %	Total	% of Variance	Cumulative %
1	5.187	28.817	28.817	3.727	20.707	20.707
2	2.241	12.453	41.270	2.523	14.019	34.726
3	1.605	8.914	50.184	2.204	12.244	46.970
4	1.277	7.097	57.281	1.597	8.874	55.844
5	1.115	6.194	63.475	1.374	7.632	63.475
6	0.974	5.413	68.888			
7	0.838	4.655	73.543			
8	0.770	4.277	77.821			
9	0.714	3.964	81.785			
10	0.543	3.018	84.803			
11	0.491	2.727	87.529			
12	0.454	2.525	90.054			
13	0.431	2.394	92.448			
14	0.371	2.063	94.511			
15	0.311	1.730	96.241			
16	0.260	1.445	97.686			
17	0.244	1.356	99.043			
18	0.172	0.957	100.000			

Extraction Method: Principal Component Analysis.

Source: SPSS Output.

TABLE 11.4
Sample QOL Scores

Settlement Name	Highest QOL Score	Settlement Name	Middle QOL Score	Settlement Name	Lowest QOL Score
Kozhikode	99.99	Cherpu	28.0188387	Nemom	3.77523941
Edapally	87.65	Areekode	27.9698945	Kuzhalmannam	3.09027226
Thiruvananthapuram	67.83	Koovappady	27.7524737	Vellanad	2.82314276
Chalakudy	66.10	Adimali	27.3923102	Alathur	0.89580821
Mulamthuruthy	60.21	Irikkur	26.9305622	Malampuzha	0.00740691

relevant. Five principal components were identified based on the criteria that their Eigen values were equal to or greater than unity, as shown in Table 11.3 (principal components).

All the 18 variables were now grouped under these five components. Thus, the dimensions of the data set were reduced from 18 to five. Squared factor loadings of the principal components were obtained from the factor analysis to convert negative values into positive. Variables with significant factor loadings (in this case >0.2) were then assigned to each of the components. Weightage of principal components was calculated by dividing the percentage variance of each component by cumulative percentage of variance. In this case, the cumulative percentage of variance was 63.475 which is a reasonably good value (Erik and Marko 2011). Weights of each component were obtained as,

WC1 = 0.32623 WC2 = 0.220852, WC3 = 0.192891, WC4 = 0.139805 and WC5 = 0.12023[2](Babu 2019).

Once the weights were found, the index scores were calculated for each settlement using the formula mentioned before. These were then transformed into standard scores through min-max transformation for better interpretation of results. A sample of 15 scores (highest 5, middle 5 and lowest 5) of the 151 units (Block panchayaths) are shown in Table 11.4.

The final scores were validated quantitatively using the cross-validation technique. It was found that The Pearson's coefficient for both the training set and test set was almost equal making the model statistically valid (refer Section 11.3.2.3). For qualitative validation, characteristics of settlements were analysed based on their known values of corresponding variables. It was observed that settlements having good housing and civic infrastructure scored high and vice versa. This validates the model as the status of settlements was linked to their actual conditions of living (Zehba 2018; Babu 2019).

11.5   DISCUSSIONS

The composite index was calculated for all the 151 block panchayats in Kerala with scores ranging from 0 to 100 based on their relative performance. Among all, block panchayats, Kozhikode secured the highest rank and Malampuzha the least. The results were then spatially linked to GIS map of their respective block panchayats and a spatial distribution map was prepared which is shown below (Figure 11.4). Hotspot analysis tool in GIS was used for the same.

As seen in the map, QOL distribution shows significant disparities among the settlements in Kerala. This variation does not follow a definitive pattern but shows much relative to the state's urbanization trends. The four highest QOL clusters are located in Kannur, Kozhikode, Eranakulam and Thiruvananthapuram districts which are distanced apart approximately equal. It is clear from the map that factors such as geography, urbanization, etc. have influenced the QOL pattern. These are discussed further.

11.5.1   SPATIAL INTERPRETATIONS AND INFERENCES

One of the important goals of this study was to analyse QOL spatially and understand the geographical correlations of QOL with other spatial characteristics. For this, the QOL scores were mapped and zones of higher and lower QOL were identified. As explained in the methodology (Section 11.3.3), three zones of high, medium and low

---

[2] WC – Weight of Component

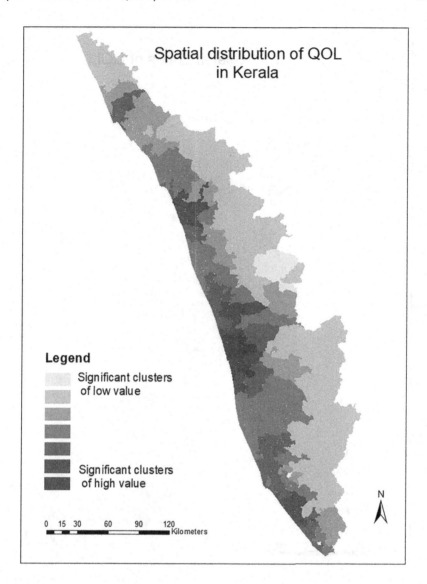

FIGURE 11.4   Spatial distribution of QOL.

A map of Kerala with heading 'Spatial distribution of QOL in Kerala' showing variation of QOL in the settlements in different shades.

QOL were calculated by dividing the standard deviation of the scores equally. Map (Figure 11.5) shows the three spatial zones of settlements.

From the map (refer Figure 11.5), it can be seen that high QOL zone was mainly located within Kozhikode, Thrissur, Eranakulam, Thiruvananthapuram and Kollam Districts of the state. These are along the coastal line of Kerala. Medium QOL was along the middle plain and low QOL zone was along the highland regions of

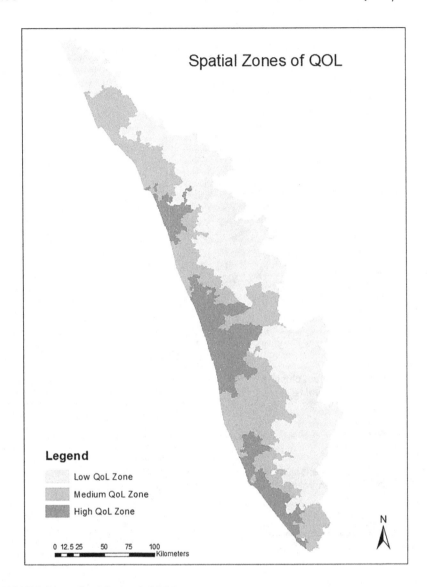

FIGURE 11.5   Spatial zones of QOL.

A map of Kerala with heading 'Spatial zones of QOL' showing low, medium and high QOL settlements in different shades with legends Low QOL, Medium QOL and High QOL.

the Western Ghats plains of Kerala (Babu 2019). The geographical factors affect a region's development considerably which in turn has an effect on its QOL pattern. This is clearly visible here.

Geographically, Kerala is divided into eastern highlands (rugged and cool mountainous terrain), central midlands (rolling hills), and western lowlands (coastal plains) (Firoz 2014; Krishnan and Firoz 2020; Kumar et al. 2019). When spatially

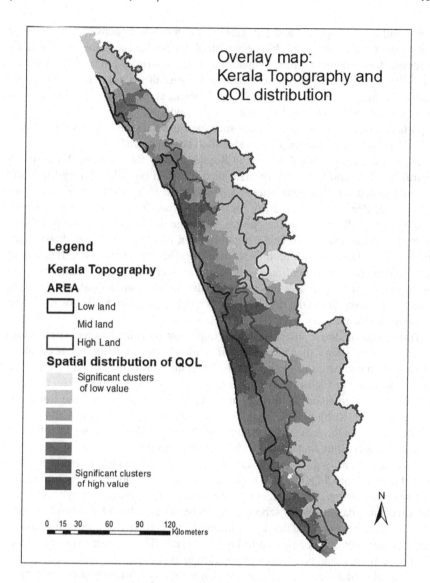

FIGURE 11.6 Topography of Kerala and QOL pattern.

A map of Kerala with heading 'Overlay map-Kerala topography and QOL distribution' showing the QOL distribution overlaid with three topographical regions with legend Lowland, Midland and Highland.

mapped (refer Figure 11.6), it was observed that settlements along the lowlands had a higher sense of QOL than those of midlands and highlands. Also, settlements along highlands scored least in QOL. This is due to the fact that it is difficult to establish infrastructure and services along high terrain areas, as compared to lowland and midland areas. As variables corresponding housing, education, health and transportation infrastructure are positively correlated with QOL index, they affect the scores significantly. Geographical and topographical constrains in construction and relatively higher vulnerability to disasters such as landslides cause the highlands less preferable for settlement when compared to midlands or lowlands. Also, lowlands have witnessed a faster pace of development because of the numerable trade opportunities offered by the coastal areas. Development created more employment and better services which eventually radiated to the midlands. Since all these factors are important variables of the QOL index, such facts have affected the results considerably. Along with this, historical facts also shall be considered.

There are three historical regions in Kerala, namely the northern 'Malabar', the central 'Cochin' and the southern 'Travancore' which prevailed during the British rule and joined together post-independence to form today's state of Kerala (Firoz 2014; Firoz, Banerji, and Sen 2016; Firoz and Kumar 2017). The characteristics of these regions still prevail in the culture, language and lifestyle of the population. From the spatial analysis of QOL, it can be seen that Cochin region shows higher number of settlements with high QOL than the other two. In Malabar region, high QOL settlements are concentrated only in and around Kozhikode which was its former commercial capital. Travancore region also shows a similar pattern. This phenomenon corresponds to the development of Cochin port and decline of Kozhikode port post-independence.

Transportation development is another important influential factor. The road network of Kerala shows that lowlands and midlands possessed higher accessibility options when compared to the highlands. Here, an overlay map of QOL scores and road network (refer Figure 11.7) reveals that both are directly linked. As explained before, the topographical constrains in laying roads have affected the transportation network in the highlands. This factor being directly connected with QOL has resulted in low QOL areas in highlands. Higher accessibility increases the efficiency of other infrastructure and services which in turn increases the overall QOL.

Population density distribution across settlements showed much similarity to the spatial distribution of QOL (refer Figure 11.8). This trend attributes back to the fact that when population density increases, the number of people served by services and infrastructure like, hospitals, schools, etc. increases and thus QOL score increases. Thus, it could be concluded that in the lowland areas, the status of infrastructure and services were better as compared to highland areas. This in turn led to people settling in such areas and the density of population increased.

It was also observed that settlements that were predominantly urban had higher QOL than those designated as rural areas (refer Figure 11.9). Urban areas by definition have higher density of population and higher percentage of working population. These factors largely influence the QOL of settlements since higher the population and population density, higher will be the population served by different infrastructure and services in a given area. In addition to this, percentage of working men is also a factor considered while calculating the QOL index. Urban areas are also perceived to have better infrastructure and services. This explained why the QOL scores of lowland urban areas were higher.

Another factor in the social equity domain is the living conditions of Scheduled Caste/Scheduled Tribe population. As shown in Figure 11.10, highlands of Kerala, especially forest areas, are the major residing place for a considerable tribal

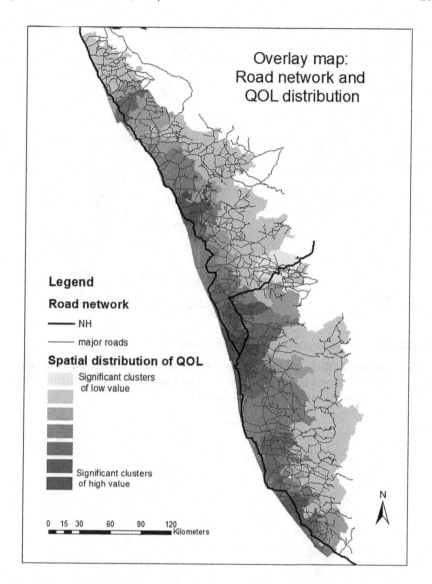

FIGURE 11.7   Road network and QOL pattern.

A map of Kerala with heading 'Overlay map-road network and QOL distribution' show-ing variation of QOL in the settlements overlaid with the road network.

population. They are generally in high deprivation conditions in terms of health and education services. This is another reason why QOL scores of highlands are low.

It was inferred from the detailed analysis of pattern of QOL that there were consid-erable disparities/inequities in living quality among the settlements of the study area despite high HDI scores. The results can be further analysed in detail with respect to each variable so as to understand how it affects the overall QOL. A QOL assessment

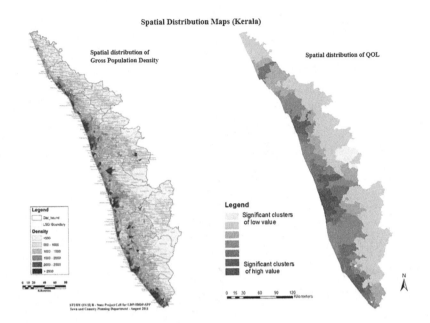

FIGURE 11.8   Comparison of population density and QOL distribution.

Two maps of Kerala, first with heading 'spatial distribution of gross population density' showing variation of population density and second with heading 'spatial distribution of QOL' showing variation of QOL in the settlements.

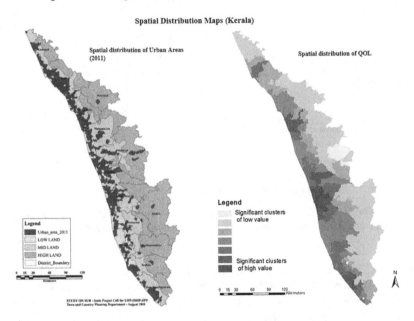

FIGURE 11.9   Comparison of urban areas and QOL distribution.

Two maps of Kerala, first with heading 'spatial distribution of urban areas' showing urban areas of Kerala and second with heading 'spatial distribution of QOL' showing variation of QOL in the settlements.

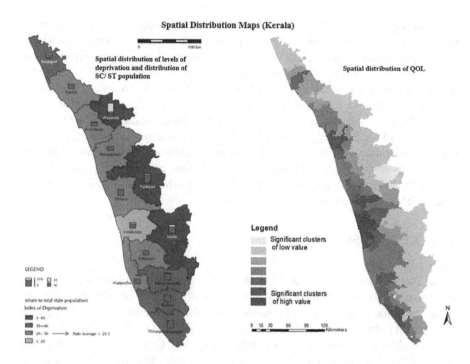

FIGURE 11.10   Comparison of SC/ST population deprivation and QOL distribution.

Two maps of Kerala, first with heading 'spatial distribution of levels of deprivation and distribution of Scheduled Cast/Scheduled Tribe (SC/ST) population' showing SC/ST population deprived settlements of Kerala and second with heading 'spatial distribution of QOL' showing variation of QOL in the settlements.

in any settlement helps in discovering the stressed areas which need intervention. The spatial distribution of QOL from the study shows a similar pattern as that of district-wise spatial distribution pattern of HDI in Kerala. District-wise distribution of schools and hospitals in Kerala also follows the same pattern. The settlements having higher values for number of schools and hospitals show high deprivation rates of SC/ST families. This might be one of the reasons why those settlements have lower scores of QOL. The parameters considered for finding the deprivation of SC/ST communities are housing quality, access to drinking water, good sanitation and electricity lighting which is almost similar to the parameters used in the study for calculating the social equality. Hence, the spatial trend derived from the study is almost similar to the already-known patterns of social indicators and variables. Thus, it can be concluded that the results obtained from the study is logically valid.

After analysis, broad planning recommendations were suggested to dissolve these disparities and improve the living conditions of the residents which are explained further. Community participation should be enhanced in the decision-making process and thereby better social characteristics of the population should be utilized. More income generating employment opportunities should be created that would benefit the educated population and thus improve the workforce participation. Skill development programmes should be expanded for creating more employment opportunities. High-income population should be encouraged in investments beneficial to infrastructure development. Community should be made responsible for improvement and mainte-nance of infrastructure. Policies should be introduced for the improvement of working conditions of the work force. Health initiatives should be introduced for improving the physical health of the population by utilizing the available good environment.

The overall methodology adopted here for QOL index construction and analysis can be used universally. However, selection of variables and index construction are the two important steps. Out of these, the latter follows a standard method. Though variables may change according to the context, the method used can be applied any-where with suitable context-based modifications.

11.6   CONCLUSIONS

The study attempted a spatial assessment of QOL in the southern state of Kerala, India. The analysis was conducted in two stages – construction of a composite index and spatial analysis. The initial task of selection of variables of QOL was done by screening a master list of variables formed form the holistic literature with the help of an expert opinion survey. Then, a composite index of QOL was constructed using these variables and corresponding secondary data. Once QOL scores were calculated for all the settlements, spatial analysis was carried out. Index scores were spatially mapped to analyse the pattern of QOL. Following are the observations from the study:

- There is an evident spatial disparity of QOL across Kerala. The quality of infrastructure, services, etc. determines the QOL within settlements.
- Geographical and topographical characters influence QOL. QOL decreases from lowlands to highlands.
- Higher QOL is concentrated in and around three historical regions, namely Malabar, Cochin and Travancore which are currently Kozhikode, Kochi and Thiruvananthapuram, respectively. These are also the most developed regions in the state.
- QOL is positively affected by the transportation network.
- QOL is high in densely populated areas.
- QOL is higher in urban areas than that in rural areas.

Finally, broad planning recommendations were stated for the improvement of state of QOL in the concerned site.

11.6.1   LIMITATIONS AND SCOPE FOR FURTHER RESEARCH

The spatial variation in QOL is a relative depiction of QOL conditions within Kerala which does not mean that areas having low scores are extremely backward in prog-ress indicators. It should be noted that the study was performed based on secondary surveys and primary data were omitted. Also, the study was limited to block pan-chayat level and was not calculated and compared with other levels of administration.

The study can be extended to lower levels of settlements with comparative analysis across settlements. Also, subjective indicators from primary data sources have vast research scopes and can be explored further in terms of assessment methods and variables. The methodological framework used here is a major contribution in any research regarding QOL. This can be used elsewhere with appropriate changes in the variable selection.

To conclude, the research is only an earnest step in analysing QOL across settlements using spatial analysis so as to give appropriate policy directions for governance, planning and management in one of the most socially advanced states of India.

REFERENCES

Askam, A., and Corrado, L. 2012. "The geography of well-being." *Journal of Economic Geography* 12: 627–649.

Babu, N. 2019. "Spatial Assessment of Well-Being and Its Determinants: A Case Study of Kerala, India." National Institute of Technology, Calicut.

Bardhan, R., Kurisu, K.H., and Hanaki, K. 2011. "Linking Urban Form & Quality of Life in Kolkata, India." In *47th ISOCARP Congress 2011*, 1–12, Kolkata.

Becker, William, Michaela Saisana, Paolo Paruolo, and Ine Vandecasteele. 2017. "Weights and Importance in Composite Indicators: Closing the Gap." *Ecological Indicators* 80 (September): 12–22. Doi: 10.1016/j.ecolind.2017.03.056.

Bogdanov, N., Meredith, D., and Efstratoglou, S. 2008. "A Typology of Rural Areas in Serbia." *Economic Annals* 53 (April). Doi: 10.2298/EKA0877007B.

Chaudhuri, P. 1977. "Poverty, Unemployment and Development Policy: A Case Study of Selected Issues with Reference to Kerala." *The Economic Journal* 87 (348): 818–819. https://doi.org/10.2307/2231394.

Chi, G., and Ventura, S. J. 2011. "Population Change and Its Driving Factors in Rural, Suburban, and Urban Areas of Wisconsin, USA, 1970–2000." Edited by Jacques Poot. *International Journal of Population Research* 2011: 856534. Doi: 10.1155/2011/856534.

Damen, R. G. 2014. *Evaluating Urban Quality and Sustainability.* University of Twente.

Delsante, I. 2016. "Urban environment quality assessment using a methodology and set of indicators for medium-density neighbourhoods: A comparative case study of Lodi and Genoa." *Ambiente Construído* 16 (3): 7–22, Doi: 10.1590/s1678-86212016000300089.

Dutt, A. K., Monroe, C. B., and Vakamudi, R. 1986. "Rural-Urban Correlates for Indian Urbanization." *Geographical Review* 76 (2): 173–83.

El-Farouk, A. 2018. "Measuring Social Progress in a Rural Setting." *International Journal of Current Research* 10 (January): 64326–34.

Erik, M., and Marko, S. 2011. *A Concise Guide to Market Research The Process, Data, and Methods Using IBM SPSS Statistics.* New York: Springer.

EUROSTAT. 2015. "Quality of life facts and views." Luxembourg.

Firoz, M.C. 2014. *Reclassification of the Typology and Pattern of Composite Settlements: A Case of Kerala, India.* (Unpublished Doctoral Thesis) Kharagpur: Indian Institute of Technology.

Firoz, M.C., Banerji, H. and Sen, J. 2016. "A methodology for defining the typology of the rural urban continuum settlments in Kerala, India." *Journal of Regional Development and Planning* 3 (1): 49–60.

Firoz, M.C., and Kumar, T.M.V. 2017. "Transforming economy of Calicut to smart economy." In *Smart Economy in Smart Cities. Advances in 21st Century Human Settlements,* 331–358. Singapore: Springer.

Franke, R., and Chasin, B. 1992. "Kerala State, India: Radical Reform as Development." *International Journal of Health Services : Planning, Administration, Evaluation* 22 (February): 139–56. Doi: 10.14452/MR-042-08-1991-01_1.

Freimann, A. 2016. "Limitations of the GDP as a measure of progress and well-being." *Ekonomski Vjesnik* XXIX 29: 257–272.

George, K.K., and Kumar, N.A. 1997. "Kerala- The land of development paradoxes." 2. CSES Working Paper. Kochi.

Ilic, K. A. 2013. *Evaluating Disparities in Quality of Life in the City of Atlanta Using an Urban Health Index.* Georgia State University.

Janaagraha. 2013. "Ward Quality Score Databook." Bengaluru.

Kallingal, F. R. 2017. "Regional Disparities in the Social Development of Kerala: A Spatial Assessment." National institute of Technology, Calicut.

Kallingal, F.R., and Firoz, M.C. 2017. "Composite index of social development: A case of Kerala, India." In *Proceedings of the International Conference on Working Class Districts/ Urban Transformations and Qualities of Life in the Growing City,* Vienna, Austria, 46–52.

Kapoor A., Kapoor, M. and Krylova, P. 2016. *Social Progress Index: States of India.* Washington, DC: Social Progress Imperative.

Krishnan, V.S., and Firoz, M.C. 2020. "Regional Urban environmental quality assessment and spatial analysis." *Journal of Urban Management* 9 (October 2019): 191–204. Doi: 10.1016/j.jum.2020.03.001.

Kumar, T.M.V., Firoz, M.C., Puthuvayi, B., Harikumar P., and Sankaran, P. 2019. "Smart water management for smart Kozhikode metropolitan area." In *Smart Environment for Smart Cities,* 241–306. Singapore: Springer.

Madu, I.A. 2010. "The structure and pattern of rurality in Nigeria." *Geojournal* 75: 175–184.

Martínez, J. 2009. "The use of GIS and indicators to monitor intra-urban inequalities. A case study in Rosario, Argentina." *Habitat International* 33 (4): 387–396. Doi: 10.1016/j. habitatint.2008.12.003.

Maynooth, N. U. I., and Martin, B. S. 2000. "Irish Rural Structure and Gaeltacht Areas."

Mazziotta, M., and Pareto, A. 2013. "Methods for constructing composite indices : One for all or all for one ? 1." *Rivista Italiana Di Economia Demografia e Statistica* LXVII 67 (2): 67–80.

Mehdi, T. 2019. "Stochastic dominance approach to OECD's better life index." *Social Indicators Research* 143: 917–954.

Michalos, A.C., Smale, B., Labonté, R., Muharjarine, N., Scott, K., Moore, K., Swystun, L. et al. 2011. "Technical paper: Canadian index of wellbeing."

Monasterolo, I., and Coppola, N. 2011. "Mapping Serbia: More targeted rural areas for better policies." In *118th Seminar, European Association of Agricultural Economists,* Ljubljana, Slovenia.

Morris M.D. 1980. "The physical quality of life index (PQLI)." *Development Digest.* Jan 18 (1): 95–109.

MUD-GOI. n.d. "Methodology for Collection and Computation of Livability Standards in Cities." New Delhi.

Naidu, V., and Nair, M. 2007. "Development Disparity in Education Sector: An Inter District Temporal Analysis in Kerala." *Esocialsciences.Com, Working Papers.*

Nardo, M., Saisana, M. Saltelli, A., and Tarantola, S. 2005. "Tools for Composite Indicators Building."

OECD. 2011. *Compendium of OECD Well-Being Indicators.*

OECD. 2008. *Handbook on Constructing Composite Indicators Methodology and User Guide.*

Paul, G. 2017. "Urbanisation in Kerala : Current trends and key challenges for sustainability introduction" *International Journal of Research in Economics and Social Sciences (IJRESS)* 7 (2): 151–59.

Porter, M.E., Stern, and Green, M. 2017. *Social Progress Index 2017.* Social Progress Imperative. https://www.socialprogressindex.com/assets/downloads/resources/en/English-2017-Social-Progress-Index-Findings-Report_embargo-d-until-June-21-2017.pdf.

Posner, S. 2010. "Estimating the Genuine Progress Indicator (GPI) for Baltimore, MD." University of Vermont.

Risser, R., Schmeidler, K., Steg, L., Forward, S., and Steg, L. 2017. "Assessment of the quality of life in cities environmental conditions and mobility assessment of the quality" *Urbani Izziv* 17 (1): 187–93.

Saisana, M. and Philippas, D. 2012. "Sustainable society index (SSI):Taking societies' pulse along social, environmental and economic issues." *Ispra: The Joint Research Centre: EU* 32: 94–106.

Santos, L. D., and Martins, I. 2014. "Intra-Urban Disparities in the Quality of Life in the City of Porto: A Spatial Analysis Contribution." 1403. CEF.UP Working Papers. Porto.

Scott Stern, A.W. 2016. *Social Progress Index 2016: Methodological Report.* Social Progress Imperative. Washington DC.

Serag, H., Din, E., Shalaby, A., Elsayed, H., and Elariane, S.A. 2013. "Principles of urban quality of life for a neighborhood." *HBRC Journal* 9 (1): 86–92. Doi: 10.1016/j.hbrcj.2013.02.007.

Subrahmanian, K.K., and S. Prasad. 2008. *Rising Inequality with High Growth- Isn't This Trend Worrisome? Analysis of Kerala Experience.*Centre for Development Studies. Thiruvananthapuram.

Teklay, R. 2012. *Adaptation and Dissonance in Quality of Life: Indicators for Urban Planning and Policy Making MSc Thesis.* http://www.itc.nl/library/papers_2012/msc/upm/berhe.pdf.

Tharakan, P.K. M. 2006. "Kerala Model Revisited : New Problems, Fresh Challenges." 15. Thiruvananthapuram. http://www.csesindia.org/admin/modules/cms/docs/publication/15.pdf.

Tharakan, P.K.M. 2008. *When the Kerala Model of Development Is Historicised : A Chronological Perspective.* Kochi.

Young, R.D. 2008. "Quality of life indicator systems–definitions, methodologies, uses, and public policy decision making." *Ipspr.Sc.Edu*, 1–19. http://www.ipspr.sc.edu/publication/Quality of Life.pdf.

Zachariah, K.C., and Rajan, I. 2014. "Dynamics of Emigration and Remittances in Kerala: Results from the Kerala Migration Survey." Thiruvananthapuram.

Zehba M. P., F. 2018. *A Spatial Assessment Framework for Evaluating Quality of Life in Urban Areas: A Case Study of Kozhikode, Kerala.* (Unpublished Master's Thesis) Calicut: National Institute of Technology.

Index

Note: **Bold** page numbers refer to tables and *Italic* page numbers refer to figures.

Printed in the United States
by Baker & Taylor Publisher Services